T0340303

Operational Excellence in the New Digital Era

Analytics and Control Series

Series Editor: Adedeji B. Badiru

Air Force Institute of Technology, Dayton, Ohio, USA

Decisions in business, industry, government, and the military are predicated on performing data analytics to generate effective and relevant decisions, which will inform appropriate control actions. The purpose of the focus series is to generate a collection of short form books focused on analytic tools and techniques for decision making and related control actions.

Mechanics of Project Management
Nuts and Bolts of Project Execution
Adedeji B. Badiru, S. Abidemi Badiru, and I. Adetokunboh Badiru

The Story of Industrial Engineering
The Rise from Shop-Floor Management to Modern Digital Engineering
Adedeji B. Badiru

Innovation
A Systems Approach
Adedeji B. Badiru

Project Management Essentials
Analytics for Control
Adedeji B. Badiru

Sustainability
Systems Engineering Approach to the Global Grand Challenge
Adedeji B. Badiru and Tina Agustiady

Operational Excellence in the New Digital Era
Adedeji B. Badiru and Lauralee Cromarty

For more information on this series, please visit: https://www.routledge.com/Analytics-and-Control/book-series/CRCAC

Operational Excellence in the New Digital Era

Adedeji B. Badiru and Lauralee Cromarty

CRC Press is an imprint of the
Taylor & Francis Group, an **informa** business

First edition published 2021
by CRC Press
6000 Broken Sound Parkway NW, Suite 300, Boca Raton, FL 33487-2742

and by CRC Press
2 Park Square, Milton Park, Abingdon, Oxon, OX14 4RN

ISBN: 9780367509811 (hbk)
ISBN: 9780367509828 (pbk)
ISBN: 9781003052036 (ebk)

DOI: 10.1201/9781003052036

Typeset in Times
by KnowledgeWorks Global Ltd.

Dedicated to the labor of love for operational excellence.

Contents

Preface

Embracing the principle that it is a systems world today, this book uses a systems-based approach to show how operational excellence can be pursued, achieved, and sustained. The systems approach facilitates process design, evaluation, justification, and integration. A key power of the book is that it explicitly highlights the role of integration in every pursuit of operational excellence, using the tools and techniques of industrial engineering. Operational excellence is the lifeline of economic development. The business climate improves for everyone when operational excellence is instituted in an organization. This book presents four compact chapters that provide techniques and methodologies for embracing and utilizing industry's best practices. Emphasis is placed on the context related to the theme of systems thinking. Some definitions are used within the framework of conceptual processes while some are used within the platform of digital technology. An organization can have operational excellence in process-related activities without being involved in technological innovation. Foremost in the process of operational excellence is the role of project management. This book offers experience-based guidelines for operational excellence. The four chapters discuss the classical principles of operations management, digital platforms for operational excellence, systems thinking in the digital era, and work performance via project management. The appendix at the end of the book presents a glossary of terminologies in the digital era.

MATLAB® is a registered trademark of The Math Works, Inc. For product information, please contact:

The Math Works, Inc.
3 Apple Hill Drive
Natick, MA 01760-2098
Tel: 508-647-7000
Fax: 508-647-7001
E-mail: HYPERLINK "mailto:info@mathworks.com" info@mathworks.com
Web: http://www.mathworks.com

Adedeji B. Badiru
Lauralee Cromarty

Preface

Case Study of Operational Excellence

AFIT wins Organizational Excellence Award from the USA Air Force

This brief case study presents a justification of why writing this book is important, useful, and timely. While this manuscript was still in production, the Air Force Institute of Technology (AFIT), where the first author, Adedeji Badiru, works, was notified that the institution has won the 2019/2020 **Organizational Excellence Award** from the USA Air Force. This means that throughout the entire USA Air Force, consisting of over 490,000 personnel (military and civilians) and hundreds of air force units, AFIT was the one that won the award for that period. This book's co-author, Adedeji Badiru, as the Dean of AFIT's graduate school and a member of the leadership team that paved the way for this award, contributed directly to this organizational accomplishment. The institution's nomination and award were based on the following ideals of an excellent organization and operational excellence:

- Leveraging digital tools and systems thinking for educational operations
- Providing advanced education for US Air Force and Space Force
- Enmeshing teaching and research missions, strategically
- Instituting consistency of academic operations
- Leveraging teamwork across the spectrum of operations
- Developing and implementing strategies for efficiency

- Sustaining effectiveness and continuity, even during the COVID-19 pandemic
- Providing multi-lateral support for other air force organizations and international allies

All of these items align directly with the strategies, tools, and techniques covered in this book. Excellent organizations thrive on operational excellence. This gives further credence to the premise of this book entitled "Operational Excellence in the New Digital Era." Thus, the tools and techniques presented in this book do work, if implemented strategically. Congratulations to AFIT!

In a similar vein of practicing what is preached for operational excellence, in 2013, Adedeji Badiru, as a department head then, was the department-level Team Leader for the Air University level award for C3 (Cost Conscious Culture), which was a phase of the Air Force Organizational Excellence recognition.

About the Authors

Adedeji B. Badiru is a professor of Systems Engineering at the Air Force Institute of Technology (AFIT). He is a registered professional engineer and a fellow of the Institute of Industrial Engineers as well as a fellow of the Nigerian Academy of Engineering. He has a BS degree in Industrial Engineering, MS in Mathematics, and MS in Industrial Engineering from Tennessee University, and PhD in Industrial Engineering from the University of Central Florida. He is the author of several books and technical journal articles and has received several awards and recognitions for his accomplishments.

Lauralee Cromarty has over 40 years' experience in manufacturing, retail, and service industries. She has worked on continuous improvement projects, managed operations, and facilitated change. She is a fellow of the Institute of Industrial and Systems Engineers. Lauralee is currently retired and a small business mentor for SCORE Austin Chapter.

Classical Principles of Operations Management

1

INTRODUCTION

The classical principles of operations management provide the foundational framework for the theme of this book on operational excellence. Every organization desires to achieve and sustain operational excellence as the basis for organizational survival. Unfortunately, not every organization has the wherewithal to perform at the level of desired excellence. The discipline of industrial engineering, because of its versatility and robust coverage of both qualitative and quantitative aspects, has been proven to be a good avenue to achieving operational excellence. In this chapter, we discuss the classical principles of operations management, which morphed into the concept of scientific management, which later formed the emergence of industrial engineering. The structure of the content of this book is based on the progressive building blocks below:

1. Foundation from the classical principles of operations management.
2. How operational excellence has changed in the digital age.
3. Systems thinking as the basis for operational excellence.
4. Teamwork and work simplification for operational excellence.

DOI: 10.1201/9781003052036-1

THE EMERGENCE OF INDUSTRIAL ENGINEERING

Industrial engineering emerged out of industry's need for efficient work systems and processes for better utilization of workers in factory and construction operations in 1900 (Badiru, 2019). One common definition of an industrial engineer states that:

> Industrial Engineer: One who is concerned with the design, installation, and improvement of integrated systems of people, materials, information, equipment, and energy by drawing upon specialized knowledge and skills in the mathematical, physical, and social sciences, together with the principles and methods of engineering analysis and design to specify, predict, and evaluate the results to be obtained from such systems.

This definition generically embodies the various aspects of what an industrial engineer does. Whether a definition is official or not, the description of the profession is always evolving as new application opportunities evolve.

An historical summary of industrial engineering reveals a great legacy of contributions to industrial and economic development. In the early history of the United States, several efforts emerged to form the future of the industrial engineering profession. George Washington was said to have been fascinated by the design of farm implements on his farm in Mount Vernon, Virginia. He had an English manufacturer send him a plow built to his specifications that included a mold on which to form new irons when old ones were worn out or would need repairs. This represents one of the early attempts to create a process of achieving a system of interchangeable parts. Thomas Jefferson invented a wooden moldboard that, when fastened to a plow, minimized the force required to pull the plow at various working depths. Jefferson also invented a device that allowed a farmer to seed four rows at a time. In pursuit of higher productivity and operational efficiency, he invented a horse-drawn threshing machine that did the work of ten men.

In 1798, Eli Whitney used mass production techniques to produce muskets for the US Army. He developed the idea of having machines make each musket part so that it could be interchangeable with other similar parts. By 1850, the principle of interchangeable parts was widely adopted. It eventually became the basis for modern mass production for assembly lines. These are examples of ingenuity and innovation in early agricultural industry. It is believed that Eli Whitney's principle of interchangeable parts contributed significantly to the Union victory during the US Civil War. Thus, the early practice of industrial engineering made significant contribution to the military and industry, even though the formal name of the profession did not emerge until much later. The heritage of industrial engineering in the military has continued until today and should be extended in a contemporary framework to meet current organizational needs for operational excellence in other spheres of organizational pursuits.

OPERATIONS MANAGEMENT AND INDUSTRIAL ENGINEERING

The definition of industrial engineering embodies the various aspects of what every organization, military or civilian, faces in terms of operational challenges. Industrial engineering is versatile, flexible, adaptive, and diverse. Industry, service, and supply chain companies have used it successfully for many decades. Because of its versatility and robustness, the military has called upon the discipline of industrial engineering for several decades to achieve program effectiveness and operational efficiencies (Badiru and Thomas, 2009). It is known from the practice of industrial engineering that a systems orientation permeates the work of industrial engineers. This is particularly applicable to all types of business and industry, where operations and functions are designed to encompass the linking of sub-systems.

Based on the foregoing examples, classical operations management forms the basis for the pursuit of operational excellence. In past decades, the classical management theory is based on the belief that workers only had physical and economic needs, with little regard for the social and job satisfaction needs of workers. Under this concept, the advocacy was for a specialization of labor, centralized leadership, top-down decision-making, and profit maximization. The emergence of industrial engineering helped to incorporate the personal needs of workers into the overall business strategies of organizations as part of operations management. Maslow's hierarchy of needs further helped to open up additional research and practice of work management systems that incorporate the human elements into the production environment. Organizations now proactively embrace these classical and scientific management theories to improve work efficiency and increase worker productivity.

Under Frederick Taylor in the early 1900s, the classical theory of management entailed a scientific study of tasks and the workers responsible for them. Although its goal was providing workers the tools necessary for maximizing their efficiency and output, it was also criticized for creating an "assembly-line" atmosphere, where employees do only menial jobs. For this reason, it was shunned and derided. Even many organizations that embraced scientific management avoided proclaiming it publicly. But the fact is that many elements of scientific management still offer good insights into how to achieve and sustain operational excellence, and some managers believe that the theory of scientific management is better suited for operations involving repetitive factory tasks.

According to Terry (2011), to understand if scientific management is suitable for an organization, it is essential to understand the foundation of classical management theory. Classical and scientific management theory is based on four main principles:

1. Company leadership should develop a standard method for doing each job using scientific management.
2. Workers should be selected for a job based on their skills and abilities.
3. Work should be planned to eliminate interruptions.
4. Wage incentives should be offered to encourage increased output.

Although the informal practice of industrial engineering has been in existence for centuries, there was no formal coalescing of the profession under one identifiable name. It has been referred to with different names and connotations. Scientific management was one of the original names used to describe what industrial engineers do. However, named or not named, humans must practice industrial engineering to achieve their goals.

Human history indicates that humans started out as nomad hunters and gatherers, drifting to wherever food could be found. About 12,000 years ago, humans learned to domesticate both plants and animals. This agricultural breakthrough allowed humans to become settlers, thereby spending less time wandering in search of food. More time was available for pursuing stable and innovative activities, which led to discoveries of better ways of planting and raising animals for food. That initial agricultural discovery eventually paved the way for the agricultural revolution. During the agricultural revolution, mechanical devices, techniques, and storage mechanisms were developed to aid the process of agriculture. These inventions made it possible for more food to be produced by fewer people. The abundance of food meant that more members of the community could spend that time for other pursuits rather than the customary labor-intensive agriculture. Naturally, these other pursuits involved the development and improvement of the tools of agriculture. The transformation from the digging stick to the metal hoe is a good example of the raw technological innovation of that time. With each technological advance, less time was required for agriculture, thereby permitting more time for further technological advancements. The advancements in agriculture slowly led to more stable settlement patterns. These patterns led to the emergence of towns and cities. With central settlements away from farmlands, there developed a need for transforming agricultural technology to domicile technology that would support the new organized community settlements. The transformed technology was later turned to other productive uses, which eventually led to the emergence of the industrial revolution. To this day, the entwined relationships between agriculture and industry can be seen. That is industrial engineering, regardless of whatever other names it is called.

Notable industrial developments that fall under the purview of the practice of industrial engineering range from the invention of the typewriter to the invention of the automobile. Writing is a basic means of communicating and preserving records. It is one of the most basic accomplishments of the society. The course of history might have taken a different path if early writing instruments had not been invented at the time they were. The initial drive to develop the typewriter was based on the need and search for more efficient and effective ways of communication. The emergence of the typewriter typifies how an industrial product might be developed through the techniques of industrial engineering. In this regard, we can consider the chronological history of the typewriter:

1714	Henry Mill obtained British patent for a writing machine. This invention opened up new avenues of documenting, disseminating, and preserving information. These capabilities positively impacted the ability to replicate and spread new operational techniques and management decisions
1833	Xavier Progin created a machine that uses separate levers for each letter. This mechanical advancement paved the way for future improvements of writing machines as new approaches to manipulating levers were developed
1843	American inventor, Charles Grover Thurber, developed a machine that moves paper horizontally to produce spacing between words and carriage return to produce spacing between lines. The crude machine was patented, but never produced for practical usage
1873	E. Remington & Sons of Ilion, NY, manufacturers of rifles and sewing machines, developed a typewriter patented by Carlos Glidden, Samuel W. Soule, and Christopher Latham Sholes, who designed the modern keyboard. This class of typewriters wrote in only uppercase letters, but contained most of the characters on the modern machines
1912	Portable typewriters were first introduced. The portability of writing machines created a new wave of mobile secretarial workforce, which proved advantageous for improving operations management

1925	Electric typewriters became popular. This made typeface more uniform. International Business Machines Corporation (IBM) was a major distributor for this product. As would be seen in decades later, electric typewriters paved the way for the origin of keyboards for computers

In each case of product development, engineers demonstrated the ability to design, develop, manufacture, implement, and improve integrated systems that include people, materials, information, equipment, energy, and other resources. Thus, product development must include an in-depth understanding of appropriate analytical, computational, experimental, implementation, and management processes and their interrelationships. That is the essence of industrial engineering.

Meanwhile in Europe, the Industrial Revolution was occurring at a rapid pace. Productivity growth, through reductions in manpower, marked the technological innovations of the 1769–1800 Europe. Sir Richard Arkwright developed a practical code of factory discipline. In their Foundry, Matthew Boulton and James Watt developed a complete and integrated engineering plant to manufacture steam engines. They developed extensive methods of market research, forecasting, plant location planning, machine layout, workflow, machine operating standards, standardization of product components, worker training, division of labor, work study, and other creative approaches to increasing productivity. This is industrial engineering! Charles Babbage, who is credited with the first idea of a computer, documented ideas on scientific methods of managing industry in his book entitled *On the Economy of Machinery and Manufacturers*, which was first published in 1832. The book contained ideas on division of labor, paying equitable wages for categories of work based on the level of importance and qualifications of workers, organization charts, and labor relations. These were all forerunners of formal industrial engineering.

The management attempts to improve productivity prior to 1880 did not consider the human element as an intrinsic factor. However, from 1880 through the first quarter of the 20th century, the works of Frederick W. Taylor, Frank and Lillian Gilbreth, and Henry L. Gantt created a long-lasting impact on productivity growth through consideration of the worker and his or her environment. This is industrial engineering!

Frederick Winslow Taylor (1856–1915) was born in the Germantown section of Philadelphia to a well-to-do family. At the age of 18, he entered the labor force, having abandoned his admission to Harvard University due to impaired vision. He became an apprentice machinist and patternmaker in a local machine shop. In 1878, when he was 22, he went to work at the Midvale Steel Works. The economy was in a depressed state at the time. Frederick was employed as a laborer. His superior intellect was very quickly recognized. He was soon advanced to the positions of time clerk, journeyman, lathe operator, gang boss, and foreman of the machine shop. By the age of 31, he was made chief engineer of the company. He attended night school and earned a degree in mechanical engineering in 1883 from Stevens Institute. As a work leader, Taylor faced the following common questions:

Which is the best way to do this job?
What should constitute a day's work?

These are still questions faced by the industrial engineers of today. Taylor set about the task of finding the proper method for doing a given piece of work, instructing the worker in following the method, maintaining standard conditions surrounding the work so that the task could be properly accomplished, and setting a definite time standard and payment of extra wages for doing the task as specified. Taylor later documented his industry management techniques in his book entitled *The Principles of Scientific Management*. This is, indeed, industrial engineering!

The work of Frank and Lillian Gilbreth coincided with the work of Frederick Taylor. In 1895, on his first day on the job as a bricklayer, Frank Gilbreth noticed that the worker assigned to teach him how to lay brick did his work three different ways. The bricklayer was insulted when Frank tried to tell him of his work inconsistencies—when training someone on the job, when performing the job himself, and when speeding up. Frank thought it was essential to find one best way to do work. Many of Frank Gilbreth's ideas were similar to Taylor's ideas. However, Gilbreth outlined procedures for analyzing each step of workflow. Gilbreth made it possible to apply science more precisely in the analysis and design of the workplace. Developing *therbligs*, which is Gilbreth spelled backward, as elemental predetermined time units, Frank and Lillian Gilbreth were able to analyze the motions of a worker in performing most factory operations in a maximum of 18 steps. Working as a team, they developed techniques that later became known as work design, methods improvement, work simplification, value engineering, and optimization. Lillian (1878–1972) brought the concern for human relations to the engineering profession. These are all industrial engineering! The foundation for establishing the profession of industrial engineering was originated by Frederick Taylor and Frank and Lillian Gilbreth. They were the first modern industrial engineers!

Henry Gantt's work advanced the management movement from an industrial management perspective. He expanded the scope of managing industrial operations. His concepts emphasized the unique needs of the worker by recommending the following considerations for managing work:

- Define his task, after a careful study
- Teach him how to do it
- Provide an incentive in terms of adequate pay or reduced hours
- Provide an incentive to surpass it

These are all steps that can be found in the practice of industrial engineering. Henry Gantt's major contribution is the Gantt chart, which went beyond the works of Frederick Taylor or the Gilbreths. The Gantt chart related every activity in the plant to the factor of time and to the other steps. This was a revolutionary concept for the time. It led to better production planning and production control. This involved visualizing the plant as a whole, as one big system made up of interrelated sub-systems. Over the past several decades, industry has been transformed from one focus level to the next, ranging from efficiency of the 1960s to the present-day trend of collaborative and digital engineering and cyber operations.

In pursuing the applications of industrial engineering, it is essential to make a distinction between the tools, techniques, models, and skills of the profession. *Tools* are the instruments, apparatus, and devices (usually visual or tangible) that are used for accomplishing an objective. *Techniques* are the means, guides, and processes for utilizing tools for accomplishing the objective. A simple and common example is the technique of using a hammer (a tool) to strike a nail to drive the nail into a wooden work piece (objective). A *model* is a bounded series of steps, principles, or procedures for accomplishing a goal. A model applied to one problem can be replicated and reapplied to other similar problems, provided the boundaries of the model fit the scope of the problem at hand. *Skills* are the human-based processes of using tools, techniques, and models to solve a variety of problems. Very important within the skills set of an industrial engineer are interpersonal skills or soft skills. This human-centric attribute of industrial engineering is what sets it apart from other engineering fields.

What follows is a chronological listing of major events that can be ascribed to the evolution and practice of industrial engineering (by whatever name it is called) over the centuries. As can be seen, industrial engineering, in any form, has been around for a long time. Wherever efficiency, effectiveness, and productivity are involved, the basis is industrial engineering. Although the listing below is not all-inclusive, it paints a picture of how the practices of industrial engineering permeate all types of operations. Knowing the history can help us trace where operational excellence originates and where it is headed in terms of our contemporary management efforts.

1440	Venetian ships are reconditioned and refitted on an assembly line
1474	Venetian Senate passes the first patent law and other industrial laws
1568	Jacques Besson publishes illustrated book on iron machinery as replacement for wooden machines
1722	René de Réaumur publishes the first handbook on iron technology
1733	John Kay patents the flying shuttle for textile manufacture—a landmark in textile mass production
1747	Jean Rodolphe Perronet establishes the first engineering school
1765	Watt invents the separate condenser, which made the steam engine the power source
1770	James Hargreaves patents his "Spinning Jenny." Jesse Ramsden devises a practical screw-cutting lathe
1774	John Wilkinson builds the first horizontal boring machine
1775	Richard Arkwright patents a mechanized mill in which raw cotton is worked into thread
1776	James Watt builds the first successful steam engine, which became a practical power source Adam Smith discusses the division of labor in *The Wealth of Nations*
1785	Edmund Cartwright patents a power loom
1793	Eli Whitney invents the "cotton gin" to separate cotton from its seeds
1797	Robert Owen uses modern labor and personnel management techniques in a spinning plant in the New Lanark Mills in Manchester, England
1798	Eli Whitney designs muskets with interchangeable parts
1801	Joseph Marie Jacquard designs automatic control for pattern-weaving looms using punched cards
1802	"Health and Morals Apprentices Act" in Britain aims at improving standards for young factory workers Marc Isambard Brunel, Samuel Benton, and Henry Maudsey designed an integrated series of 43 machines to mass-produce pulley blocks for ships
1818	Institution of Civil Engineers founded in Britain
1824	The repeal of the Combination Act in Britain legalizes trade unions
1829	Mathematician Charles Babbage designs "analytical engine," a forerunner of the modern digital computer
1831	Charles Babbage published *On the Economy of Machines and Manufacturers*
1832	The Sadler Report exposes the exploitation of workers and the brutality practiced within factories
1833	Factory law enacted in the United Kingdom. The Factory Act regulates British children's working hours A General Trades Union is formed in New York

1835 Andrew Ure published *The Philosophy of Manufacturers*, described the new industrial system that had developed in England over the course of the previous century. He advised that the new factory system is beneficial to workers because it relieved them of much of the tedium of manufacturing goods by hand

It is in this same year that Samuel Morse invents the telegraph, which opened up communication between remote locations. This invention is the forerunner of digital communication as we know it today

1845 Friederich Engels published *Condition of the Working Classes in England*, which represented a rigorous formal a study of the industrial working class in Victorian England. This same sense of paying attention to the conditions, plight, and opportunities of workers, which can still be seen in the industrial engineering of today

1847 The British Government passed the **Factory Act** to improve conditions of women and children working in factories. Young children were working very long hours in deplorable working environments. Under the Act, no child workers under nine years of age were allowed and employers must have an age certificate for their child workers. Children of 9–13 years to work no more than nine hours a day

George Stephenson founded the Institution of Mechanical Engineers. He was a British civil engineer and mechanical engineer. Renowned as the "Father of Railways," Stephenson was considered by the Victorians a great example of diligent application and thirst for improvement. It can be seen that the early works of industrial engineering mirror the early industrial focus of mechanical engineering. The "thirst for improvement" is still seen in today's industrial engineering commitment to continuous improvement

1856 Henry Bessemer revolutionizes the steel industry through a novel design for a converter

1869 Transcontinental railroad completed in the United States

1871 British Trade Unions are legalized by Act of Parliament

1876 Alexander Graham Bell invents a usable telephone. As an advancement over Samuel Morse's telegraph, the usable telephone facilitated and expedited communication beyond comprehension in that era

1877 Thomas Edison invents the phonograph. The phonograph was developed as a result of Thomas Edison's work on two other inventions, the telegraph and the telephone. In 1877, Edison was working on a machine that would transcribe telegraphic messages through indentations on paper tape, which could later be sent over the telegraph repeatedly. The recorded voice quickly became a tool for operational improvements in industrial establishments

1878 Frederick W. Taylor joins Midvale Steel Company. Taylor was widely known for his methods to improve industrial efficiency. He was one of the first management consultants. Taylor was one of the intellectual leaders of the Efficiency Movement and his ideas, broadly conceived, were highly influential in the Progressive Era of 1890s–1920s. Efficiency, in any form, whether digital or analog, has remained a focus in the practice of industrial engineering of today

1880 American Society of Mechanical Engineers (ASME) is organized

1881 Frederick Taylor begins time study experiments that would lay the foundation for the early introduction of industrial engineering to industries

1885 Frank B. Gilbreth begins motion study research. A **time and motion study** (or **time-motion study**) is an efficiency technique that combined the time study work of Frederick Taylor with the motion study work of Frank and Lillian Gilbreth. It was a major part of scientific management (also known as Taylorism). After its first introduction, time study developed in the direction of establishing standard times for work elements, while motion study evolved into a technique for improving work methods. The two techniques became integrated and refined into a widely accepted method applicable to the improvement and upgrading of work systems, which is still practiced by industrial engineering today under the name of methods engineering, which can be found in industrial establishments, service organizations, banks, schools, hospitals, government, and the military. The tools and techniques of industrial engineering are everywhere

1886 Henry R. Towne presents the paper, *The Engineer as Economist*
American Federation of Labor (AFL) is organized
Vilfredo Pareto published *Course in Political Economy*
Charles M. Hall and Paul L. Herault independently invent an inexpensive method of making aluminum

1888 Nikola Tesla invents the alternating current induction motor, enabling electricity to take over from steam as the main provider of power for industrial machines
Dr. Herman Hollerith invents the electric tabulator machine, the first successful data processing machine

1890 Sherman Anti-Trust Act is enacted in the United States

1892 Gilbreth completes motion study of bricklaying

1893 Taylor begins work as a consulting engineer

1895 Taylor presents paper entitled *A Piece-Rate System* to ASME

1898 Taylor begins time study at Bethlehem Steel
Taylor and Maunsel White develop process for heat-treating high-speed tool steels

1899 Carl G. Barth invents a slide rule for calculating metal cutting speed as part of Taylor system of management

1901	American national standards are established Yawata Steel begins operation in Japan
1903	Taylor presents paper entitled *Shop Management* to ASME H. L. Gantt develops the "Gantt Chart" Hugo Diemer writes *Factory Organization and Administration* Ford Motor Company is established
1904	Harrington Emerson implements Santa Fe Railroad improvement Thorstein B. Veblen: *The Theory of Business Enterprise*
1906	Taylor establishes metal-cutting theory for machine tools Vilfredo Pareto: *Manual of Political Economy*
1907	Gilbreth uses time study for construction
1908	Model T Ford is built Pennsylvania State College introduces the first university course in industrial engineering
1911	Taylor published *The Principles of Scientific Management* Gilbreth published *Motion Study* Factory laws are enacted in Japan
1912	Harrington Emerson published *The Twelve Principles of Efficiency* Frank and Lillian Gilbreth presented the concept of "therbligs" Yokokawa translates into Japanese Taylor's *Shop Management* and *The Principles of Scientific Management*
1913	Henry Ford establishes a plant at Highland Park, Michigan, which utilizes the principles of uniformity and interchangeability of parts and of the moving assembly line by means of conveyor belt Hugo Münsterberg published *Psychology of Industrial Efficiency*
1914	World War I starts Clarence B. Thompson edits *Scientific Management*, a collection of articles on Taylor's system of management
1915	Taylor's system is used at Niigata Engineering's Kamata plant in Japan Robert Hoxie published *Scientific Management and Labor* Lillian Gilbreth earned PhD at Brown University in 1915 in psychology
1916	Lillian Gilbreth published *The Psychology of Management* Taylor Society established in the United States
1917	The Gilbreths published *Applied Motion Study* The Society of Industrial Engineers is formed in the United States
1918	Mary P. Follett published *The New State: Group Organization, the Solution of Popular Government*
1919	Henry L. Gantt published *Organization for Work*
1920	Merrick Hathaway presents paper: *Time Study as a Basis for Rate Setting* General Electric establishes divisional organization Karel Capek: *Rossum's Universal Robots.* This play coined the word "robot"

1921 The Gilbreths introduce process analysis symbols to ASME

1922 Toyoda Sakiichi's automatic loom is developed
Henry Ford published *My Life and Work*

1924 The Gilbreths announce results of micromotion study using therbligs
Elton Mayo conducts illumination experiments at Western Electric

1926 Henry Ford published *Today and Tomorrow*

1927 Elton Mayo and others begin relay assembly test room study at the Hawthorne plant

1929 Great Depression
International Scientific Management Conference held in France

1930 Hathaway: *Machining and Standard Times*
Allan H. Mogensen discusses 11 principles for work simplification in *Work Simplification*
Henry Ford published *Moving Forward*

1931 Dr. Walter Shewhart published *Economic Control of the Quality of Manufactured Product*

1932 Aldous Huxley published *Brave New World*, the satire which prophesies a horrifying future ruled by industry

1934 General Electric performs micromotion studies

1936 The word "automation" is first used by D. S. Harder of General Motors. It is used to signify the use of transfer machines that carry parts automatically from one machine to the next, thereby linking the tools into an integrated production line
Charlie Chaplin produces *Modern Times*, a film showing an assembly line worker driven insane by routine and unrelenting pressure of his job

1937 Ralph M. Barnes: *Motion and Time Study*

1941 R. L. Morrow publishes *Ratio Delay Study*, an article in *Mechanical Engineering* journal
Fritz J. Roethlisberger: *Management and Morale*

1943 ASME work standardization committee publishes glossary of industrial engineering terms

1945 Marvin E. Mundel devises "memo-motion" study, a form of work measurement using time-lapse photography
Joseph H. Quick devises work factors (WF) method
Shigeo Shingo presents concept of production as a network of processes and operations and identifies lot delays as source of delay between processes, at a technical meeting of the Japan Management Association

1946 The first all-electronic digital computer ENIAC (Electronic Numerical Integrator and Computer) is built at Pennsylvania University
The first fully automatic system of assembly is applied at the Ford Motor Plant

1947 American mathematician, Norbert Wiener: *Cybernetics*

1948 H. B. Maynard and others introduce methods time measurement (MTM) method
Larry T. Miles develops value analysis (VA) at General Electric
Shigeo Shingo announces process-based machine layout
American Institute of Industrial Engineers is formed

1950 Marvin E. Mundel: *Motion and Time Study, Improving Productivity*

1951 Inductive statistical quality control is introduced to Japan from the United States

1952 Role and sampling study of industrial engineering conducted at ASME

1953 B. F. Skinner: *Science of Human Behavior*

1956 New definition of industrial engineering is presented at the American Institute of Industrial Engineers Convention: "Industrial engineering is concerned with the design, improvement and installation of integrated systems of people, material, information, equipment and energy. It draws upon specialized knowledge and skills in the mathematical, physical and social sciences, together with the principles and methods of engineering analysis and design to specify, predict and evaluate the results to be obtained from such systems"

1957 Chris Argyris: *Personality and Organization*
Herbert A. Simon: *Organizations*
R. L. Morrow: *Motion and Time Study*
Shigeo Shingo introduces scientific thinking mechanism (STM) for improvements
The Treaty of Rome established the European Economic Community

1960 Douglas M. McGregor: *The Human Side of Enterprise*

1961 Rensis Likert: *New Patterns of Management*
Shigeo Shingo devises the Zero Quality Control (ZQC) system (source inspection and poka-yoke systems)
Texas Instruments patents the silicon chip integrated circuit

1963 H. B. Maynard: *Industrial Engineering Handbook*
Gerald Nadler: *Work Design*

1964 Abraham Maslow: *Motivation and Personality*

1965 Transistors are fitted into miniaturized "integrated circuits"

1966 Frederick Hertzberg: *Work and the Nature of Man*

1968 Roethlisberger: *Man in Organization*
US Department of Defense: *Principles and Applications of Value Engineering*

1969 Shigeo Shingo develops single-minute exchange of dies (SMED)
Shigeo Shingo introduces pre-automation
Wickham Skinner: *Manufacturing—missing link in corporate strategy* article in *Harvard Business Review*

1971 Taiichi Ohno completes the Toyota production system
Intel Corporation develops the microprocessor chip

1973 First annual Systems Engineering Conference of AIIE

1975 Shigeo Shingo extols NSP-SS (nonstock production) system
Joseph Orlicky: *MRP: Material Requirements Planning*

1976 IBM markets the first personal computer

1980 Matsushita Electric used Mikuni method for washing machine production
Shigeo Shingo: *Study of the Toyota Production System from an Industrial Engineering Viewpoint*

1981 Oliver Wight: *Manufacturing Resource Planning: MRP II*
American Institute of Industrial Engineers (AIIE) became Institute of Industrial Engineers (IIE)

1982 Gavriel Salvendy: *Handbook of Industrial Engineering*

1984 Shigeo Shingo: *A Revolution in Manufacturing: The SMED System*
Emergence of more formal practice area named Hospital Industrial Engineering to leverage industrial engineering tools and techniques for operational improvement in the healthcare industry. National hospital operational excellence has since been credited to the application of industrial engineering

1989 Development of Code Division Multiple Access (CDMA) for Cellular Communications

1990 Wide use of the concept of Total Quality Management (TQM)

1995 The dot-com boom started in earnest
Netscape search engine was introduced
Peter Norvig and Stuart Norvig published *Artificial Intelligence: A Modern Approach*, which later became the authoritative textbook on AI

2000 The turning point of the 21st century and the Y2K computer date scare

2004 The birth of Facebook social networking
Skype took over worldwide online communication

2008 The National Academy of Engineering (NAE) published the 14 Grand Challenges for Engineering

2009 Adedeji Badiru and Marlin Thomas published the *Handbook of Military Industrial Engineering* to promote the application of industrial engineering in national defense strategies, which won the 2010 book-of-the-year award from IISE

2014 Adedeji Badiru published the second edition of the *Handbook of Industrial and Systems Engineering*
Unmanned aerial vehicles (UAV aka Drones) emerged as practical for a variety of applications

2016 Wide appearance of self-driving cars
Institute of Industrial Engineers (IIE) changed name to Institute of Industrial & Systems Engineering (IISE)

2017	Internet of Things (IoT) made a big splash
2018	Emergence of hybrid academic programs encompassing industrial engineering, digital engineering, data analytics, and virtual reality simulation
2019	Adedeji Badiru published *The Story of Industrial Engineering*, which won the 2020 book-of-the-year award from IISE
2020	Tools and techniques of industrial engineering applied to supply chain networks related to healthcare emergency needs necessitated by the COVID-19 pandemic

The nature of work has changed drastically in the new digital era (Badiru and Bommer, 2017). The change has been intensified more by the COVID-19 pandemic, with new opportunities to apply the tools and techniques of industrial engineering for operational excellence.

Additional historical and operational excellence of industrial engineering can be found in Emerson and Naehring (1988), Martin-Vega (2001), Salvendy (2001), Shtub and Cohen (2016), Sink et al. (2001), Zandin (2001), and Badiru (2014b).

PROGRESSIVE ROLE OF INDUSTRIAL ENGINEERING IN OPERATIONAL EXCELLENCE

As we move deeper into the digital era, the industrial engineering role is expanding with its growth. Figure 1.1 shows this growth. In the 1960s, coming out of the turn of the century and beyond World War I, the depression era, and the rebuilding boom of post-World War II, the overriding focus was on doing things more efficiently. The industrial engineering profession stepped forward to take the lead in the efficiency drive in diverse organizational settings and operational scenarios. In the 1970s, the interest in efficiency expanded to a more explicit consciousness about product quality. This is a departure from the early classical 1909 Ford Motor era where Henry Ford was quoted as saying that "Any customer can have a car painted any color that he wants so long as it is black." The reason for such a policy was the belief that having only one color would lead to production efficiencies, which would lead to lower costs of production, which would lead to lower prices for consumers.

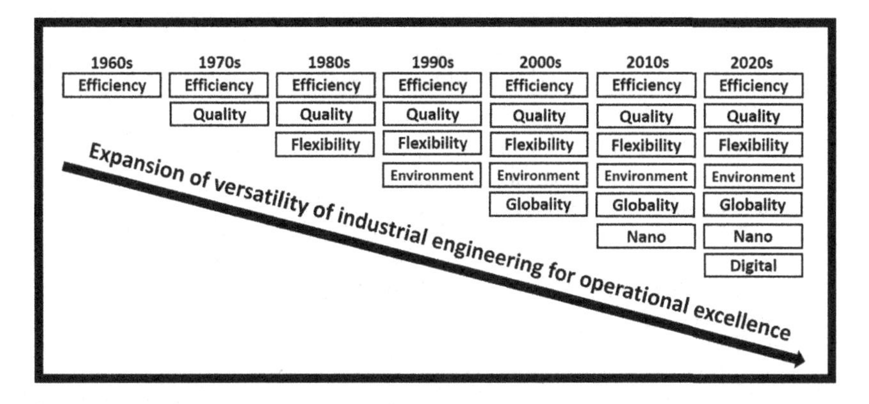

FIGURE 1.1 Expanding roles, relevance, and applicability of industrial engineering for operational excellence.

The quality consciousness drove the development of a variety of new methodologies for quality management (Badiru, 2014a). As usual, industrial engineering stepped forward to take the lead in developing new quality control techniques, based on statistically rigorous processes. In the 1980s, flexibility in production systems was added to the expanding scope of interests in business and industry. With carefully designed flexibility strategies, organizations could adapt and be more responsive to market demands. In the 1990s, environmental consciousness became a major interest. Business, industry, and academia worked more collaboratively to attend to environmental concerns. Again, industrial engineering was in the mix of the environmental impact assessments. In the 2000s, the expanded modes of global travel and communication led to globality being of interest. In 2010s, business and industry demonstrated that if you look at the big scope (global), you must also look at the small items. Thus, the movement toward nanoscience developed. This now brings us to what is trending today, along the line of the digital era. In each case, the flexibility, adaptability, and versatility of the profession of industrial engineering provide a sustainable pathway to operational excellence, regardless of whatever is trending in business and industry. The digital era offers more opportunities for operational excellence, in the manner of what industrial engineering has provided for decades on non-digital platforms.

As demonstrated in the progression of increasing operational responsibilities, industrial engineering is methodologically based and a robust discipline of engineering. It provides a framework to focus on whatever area of interest, and incorporates inputs from a variety of disciplines while maintaining the engineer's familiarity and grasp of physical processes.

REFERENCES

Badiru, Adedeji B. (2014a), "Quality Insights: The DEJI® Model for Quality Design, Evaluation, Justification, and Integration," *International Journal of Quality Engineering and Technology*, Vol. 4, No. 4, pp. 369–378.

Badiru, Adedeji B. (2014b), editor, *Handbook of Industrial & Systems Engineering*, 2nd edition, Taylor & Francis Group/CRC Press, Boca Raton, FL.

Badiru, Adedeji B. (2019), *The Story of Industrial Engineering: The Rise from Shop-Floor Management to Modern Digital Engineering*, Taylor & Francis Group/CRC Press, Boca Raton, FL.

Badiru, Adedeji B. and Marlin U. Thomas (2009), editors, *Handbook of Military Industrial Engineering*, Taylor & Francis Group/CRC Press, Boca Raton, FL.

Badiru, Adedeji B. and Sharon C. Bommer (2017), *Work Design: A Systematic Approach*, Taylor & Francis Group/CRC Press, Boca Raton, FL.

Emerson, Howard P. and Douglas C. E. Naehring (1988), *Origins of Industrial Engineering: The Early Years of a Profession*, Industrial Engineering & Management Press, Institute of Industrial Engineers, Norcross, GA.

Martin-Vega, Louis A. (2001), "The Purpose and Evolution of Industrial Engineering," in Zandin, Kjell B., editor, *Maynard's Industrial Engineering Handbook*, 5th edition, McGraw-Hill, New York, NY.

Salvendy, Gavriel (2001), editor, *Handbook of Industrial Engineering: Technology and Operations Management*, 3rd edition, John Wiley & Sons, New York, NY.

Shtub, Avraham and Yuval Cohen (2016), *Introduction to Industrial Engineering*, 2nd edition, Taylor & Francis Group/CRC Press, Boca Raton, FL.

Sink, D. Scott, David F. Poirier and George L. Smith (2001), "Full Potential Utilization of Industrial and Systems Engineering in Organizations," in Salvendy, Gavriel, editor, *Handbook of Industrial Engineering: Technology and Operations Management*, 3rd edition, John Wiley & Sons, New York, NY.

Terry, Lea (2011), "Classical and Scientific Management Theory," https://www.business.com/articles/classical-and-scientific-management-theory/, accessed February 2, 2021.

Zandin, Kjell B. (2001), editor, *Maynard's Industrial Engineering Handbook*, 5th edition, McGraw-Hill, New York, NY.

Digital Platforms for Operational Excellence

2

DIGITAL BACKBONE FOR OPERATIONAL EXCELLENCE

In this chapter, we discuss how operational excellence has changed in the digital era. Many of the discussions are in the context of the concepts, tools, and techniques of industrial engineering, which is a profession that embraces and leverages new ways of doing things. Over the course of its evolution into a separate and distinct discipline, industrial engineering has always been one of the first disciplines to identify, embrace, and capitalize on new and emerging technologies. In this respect, the digital era presents a unique new environment to bring the ideals of industrial engineering to bear on operational excellence. As chronologically presented in Chapter 1, industrial engineering has always progressed with new technologies and operational excellence opportunities. From an operational perspective, the digital era consists primarily of digital-based science, technology, engineering, and mathematics (STEM). In this chapter, the focus is on the tools and techniques of digital engineering from the lens of industrial engineering.

DOI: 10.1201/9781003052036-2

DIGITAL PLATFORM FOR MANUSCRIPT ENHANCEMENT

A good example of how the digital era has facilitated operational excellence can be seen in the processing of book manuscripts in the modern era, as in the case of this book. In the past, manuscript interfaces between the author and the publisher took place through multiple shipments of boxes of the printed manuscript back and forth between the author, the acquisition editor, the copyeditor, the cover designer, and others in the book production process. Nowadays, all the intermediate activities and interfaces are done electronically. Even the book agreement signing process is now done via digital document signing tools. The digital platform has improved book publishing operations. The consequence is better quantity and higher quality of book publishing outputs. That is a good testimony for operational excellence.

NEXUS OF DIGITAL ENGINEERING AND INDUSTRIAL ENGINEERING

To utilize the framework of industrial engineering to advance the applications of digital engineering, we must have a common understanding of what digital engineering entails. Otherwise, we would be groping in the dark digital abyss of a new hype. For our common understanding, ***digital engineering*** is the combined art and science of creating, capturing, designing, evaluating, justifying, and integrating data using digital (i.e., electronic) tools and processes. This requires the humans in the loop of the process to also have a digital mindset. A digital tool that is devoid of the digital readiness of humans will not work. So, workforce development along the digital spectrum is essential for a sustainable success. Overlaying the above definition of digital engineering on the common definition of industrial engineering (from Chapter 1), we see a commonality in how industrial engineering directly fits the expectations, goals, and objectives of the digital era. For topical relevance in this section, the definition of industrial engineering is re-emphasized below:

> Industrial Engineering is concerned with the design, installation, and improvement of integrated systems of people, materials, information, equipment, and energy by drawing upon specialized knowledge and skills in the

> mathematical, physical, and social sciences, together with the principles and methods of engineering analysis and design to specify, predict, and evaluate the results to be obtained from such systems.

The above is exactly what the digital era calls for. Operational excellence cannot evolve and survive only on the basis of technological assets. Other assets, particularly human-centric resources must be factored into the digital era. Therefore, industrial engineering is of vital relevance for operational excellence in a digital era.

What is operational excellence? Operational excellence infers accomplishing work in the best possible way. So, operational excellence requires using industrial engineering methodologies.

NATIONAL DEFENSE PUSH FOR DIGITAL ENGINEERING

A common saying goes something like "As the military goes, so goes the nation." National defense needs have fueled more interest in the digital era. Much is being said today about the emergence of digital engineering in defense operations. Yes, indeed, digital engineering promises to revolutionize operations that had, hitherto, been tackled with analog processes—and a lot of time and resources. The US Air Force now has a digital engineering roadmap that business and industry can adapt and utilize to its needs. In July 2018, the US Department of Defense (DoD) released its Digital Engineering Strategy for the purpose of promoting the use of digital representations of systems and components and using digital artifacts to design and sustain national defense systems. The pursuit of this strategy is now spreading rapidly throughout all segments of the DoD. Fortunately, industrial engineering is already aligned with the digital era, in that, it has always sought methods and technologies to solve problems and incorporate them into the industrial engineering toolbox. So, industrial engineering is well-positioned and prepared to offer guidelines and implementation strategies for operational excellence in the digital era. The premise of this book is exactly of that motive. Industrial engineering can be the driver for the multitude of topics embedded within digital engineering. This can be done through partnerships and collaborations with like-focused organizations and individuals but from a system of systems perspective. The application of a systems approach is addressed in the next chapter. Far too often, new strategic pursuits are embraced from disjointed approaches. A systems framework can ensure that all the parts are operating in consonance such that the output of the overall system would be higher than the mere sum of the individual outputs of disconnected subsystems. In this regard, integration of efforts in the digital era is the key to success. As a society embracing the digital era, we must fundamentally change the way we manage, monitor, and control operations. Business, industry, academia, and government must collectively be ready to embrace industrial and systems engineering methodologies.

WORLDWIDE PUSH FOR OPERATIONAL EXCELLENCE

Operational excellence is a worldwide requirement in today's fast-paced business setting, which all organizations must spire to achieve. Joseph F. Paris, Jr., Chairman of XONITEK Group of Companies explains operational excellence as "a state of readiness that is attained as the efforts throughout the organization reach a state of alignment for achieving its strategies; and where the corporate culture is committed to the continuous and deliberate improvement of company performance and the circumstances of those who work there—to pursue 'Operational Excellence by Design,' and not by coincidence."

In essence, any organization that can take care of the various aspects touched upon in the above definition is well on its way to achieving operational excellence. Many technical and managerial topics are involved in the pursuit of operational excellence. Most of these are within the purview of the industrial engineering profession. Some of the common tools and techniques of industrial engineering include Lean business strategy, Six Sigma, quality management, efficiency analysis, productivity improvement, and project management. Some of these topics are covered in this chapter while others are covered in subsequent chapters. Added to these, and often connecting these tools, is a plethora of digital tools such as the Internet and WIFI, faster and more powerful computers, new fast, powerful software for analysis and modeling, artificial intelligence (AI), big data, bigger and bigger monitors for displaying data, collaboration software, powerful handheld devices to put computing at the point of work, project management software, radio frequency identification (RFID) technology, miniature sensors and probes, and many more.

INPUT-PROCESS-OUTPUT FRAMEWORK

A systems approach to operational excellence sees a business as having three parts: inputs, processes, and outputs. Digital communication is the backbone of controlling inputs, connecting the processes, and monitoring outputs to promote fast, responsive actions to achieve operational excellence. The digital era has vastly improved communications in all three parts, thus speeding up the delivery of products and services tremendously, improving quality, and consequently increasing the ability to produce custom products at the same speeds and quality. If the different elements of the organization are integrated as a cohesive enterprise, communication can lead to higher efficiencies, better operational effectiveness, and cost savings. These are discussed in the framework of inputs, processes, and outputs.

Inputs

Inputs typically consist of a clear organizational vision, resources, skilled workforce, process requirements, customer desires, raw materials, market structure, and so on. Each organization must identify and define its collection of pertinent inputs, which can be variable and dynamic. Digital techniques are useful for tracking, storing, and quickly retrieving information about inputs. Data sources, digital signatures, collaboration, support, change management, organizational management system, and time-stamping of activities can be components of managing inputs in a digital environment.

Processes

Processes are the inherent capabilities of the organization to utilize the inputs to generate products, services, and/or results. Typical processes within an organization may include policies, procedures, production capabilities, design, optimization models, and so on. The digital process of an organization uses the inputs and organization strategy to transform ideas into results into outputs. Digital controls are now embedded in the processes to ensure that the end goals are achieved. Data security, credentialing, commercial external interfaces, information sharing, training, cybersecurity, outsourcing, cloud computing, and learning systems can be parts of an organization's processes, helping to monitor and coordinate the process for optimal effectiveness. For example, sensors embedded in equipment can sense out of control conditions faster than ever before and alert the operator, allowing for quick response in correcting the condition, reducing waste. Utilizing bar codes and scanners can help track raw materials, allowing for signals to be sent to the suppliers directly, reducing replenishment times and inventory levels.

Outputs

Outputs consist of physical products, desired services, and/or essential results. New inventions, a better customer experience, resilience, adaptability, concept convergence, an enhanced strategy, and so on, can be parts of an organization's expected outputs. Outputs are better achieved when they are linked to the available resources and the governing processes of the organization. Signals can be sent back to production scheduling that an item has been sold, allowing for faster replenishment; shipping can be automated to pack, signal the shipper for pickup, and track delivery, speeding up the delivery and enhancing communication to the customer.

DIGITAL DATA REQUIREMENTS FOR OPERATIONAL EXCELLENCE

Data is the basis for operational excellence. In the digital era, this is even more critical. The use of computers is the basis for all digital tools. This requires data. Every decision requires data collection, measurement, and analysis. In typical industrial engineering fashion, data may need to be collected on decision factors, costs, performance levels, outputs, and so on. In practice, we encounter different types of measurement scales depending on the particular items of interest. The different types of data measurement scales that are applicable are presented below. The different scales are important for different digital applications.

Nominal Scale of Measurement

Nominal scale is the lowest level of measurement scales. It classifies items into categories. The categories are mutually exclusive and collectively exhaustive. That is, the categories do not overlap and they cover all possible categories of the characteristics being observed. For example, in the analysis of the critical path in a project network, each job is classified as either critical or not critical. Gender, type of industry, job classification, and color are examples of measurements on a nominal scale.

Ordinal Scale of Measurement

Ordinal scale is distinguished from a nominal scale by the property of order among the categories. An example is the process of prioritizing project tasks for resource allocation. We know that first is above second, but we do not know how far above. Similarly, we know that better is preferred to good, but we do not know by how much. In quality control, the ABC classification of items based on the Pareto distribution is an example of a measurement on an ordinal scale.

Interval Scale of Measurement

Interval scale is distinguished from an ordinal scale by having equal intervals between the units of measurement. The assignment of priority ratings to project objectives on a scale of 0–10 is an example of a measurement on an interval scale. Even though an objective may have a priority rating of zero, it does not mean that the objective has absolutely no significance to the project team. Similarly, the scoring of zero on an examination does not imply that a student knows absolutely nothing about the materials covered by the examination. Temperature is a good example of an item that is measured on an interval scale. Even though there is a zero point on the temperature scale, it is an arbitrary relative measure. Other examples of interval scale are intelligence quotient (IQ) measurements and aptitude ratings.

Ratio Scale Measurement

Ratio scale has the same properties of an interval scale, but with a true zero point. For example, an estimate of zero-time unit for the duration of a task is a ratio scale measurement. Other examples of items measured on a ratio scale are cost, time, volume, length, height, weight, and inventory level. Many of the items measured in engineering systems will be on a ratio scale.

Another important aspect of measurement involves the classification scheme used. Most systems will have both quantitative and qualitative data. Quantitative data require that we describe the characteristics of the items being studied numerically. Qualitative data, on the other hand, are associated with attributes that are not measured numerically. Most items measured on the nominal and ordinal scales will normally be classified into the qualitative data category while those measured on the interval and ratio scales will normally be classified into the quantitative data category. The implication for engineering system control is that qualitative data can lead to bias in the control mechanism because qualitative data are subject to the personal views and interpretations of the person using the data. As much as possible, data for an engineering systems control should be based on a quantitative measurement. As a summary, examples of the four types of data measurement scales are presented below:

- Nominal scale (attribute of classification): Color, gender, design type
- Ordinal scale (attribute of order): First, second, low, high, good, better
- Interval scale (attribute of relative measure): IQ, grade point average, temperature
- Ratio (attribute of true zero): Cost, voltage, income, budget

Notice that temperature is included in the "relative" category rather the "true zero" category. Even though there are zero temperature points on the common temperature scales (i.e., Fahrenheit, Celsius, and Kelvin), those points are experimentally or theoretically established. They are not true points as one might find in a counting system.

DIGITAL COMMUNICATION FOR OPERATIONAL EXCELLENCE

As seen in the discussion of inputs, processes, and outputs above, communication is at the core of the digital tools. Digital communication, if structured properly, facilitates enterprise transformation, which is a strategy that encompasses services and operations that directly or indirectly impact an organization's digital experience. Communication is key to pursuing operational excellence. Communication provides the foundation for what needs to be done, why, when, where, how, and by whom. In the digital era, the ubiquitous availability of new communication modes, tools, and timing has enhanced communication, which, in turn, has sped up commerce. Speed is expected today—next day and now same-day delivery is becoming commonplace, and having results, knowing what is going on, on the production floor, in supply chains, and about financial transactions all now require fast communication. It is essential to leverage the new communication assets to achieve operational efficiency and effectiveness that constitute the essence of operational excellence.

Technology has changed communication in positive ways. However, there can be pitfalls in the speed and reach of digital communication. Safeguards must be put in place to ensure preservation of the integrity of communication and the information it conveys for the purpose of operational excellence. As described in the chronology of classical principles of operations management in Chapter 1, the concepts, tools, techniques, and the practice of industrial engineering have implications for operational performance. Communication is a keystone of how industrial engineers (IEs) pursue operational excellence. Today's fast-paced environment requires close, careful communication. No longer can we rely on walking over to another office or traveling to meetings to discuss issues, concepts, and operations. We cannot only view the immediate environment—we must understand the bigger system and our place in it. We must be able to measure and control our place in this bigger system, and faster, and IEs have a role to play in this as integrators and systems managers. IEs have the skills to research, learn, and incorporate these digital tools into their effort to achieve operational excellence. To convey the role of IEs in helping organizations achieve operational excellence, we provide the following succinct definition:

> Using communication linkages, industrial engineers make systems function better together with less waste, better quality, fewer resources, and on target with goals, objectives, and requirements.

Embodied in the above definition is the role of IEs as integrators and communicators. Industrial engineering will be the function that figures out how these new tools will be used in enterprises and will use these tools to achieve operational excellence.

Cooperation and coordination are paramount to a successful implementation of any improvement program. Spending more does not always translate to better outcomes. The trade-offs between time, budget, and performance must chart a path through the application of digital tools and techniques. The current era of digital operations has created better communication to improve resource utilization by providing the ability to collaborate virtually, reducing the resources needed to travel for meetings, and giving more time to personnel for value-added efforts. However, this increased speed of communication can also make the waste and ineptitude worse, faster. Some of the IE tools for digital design and analysis for improving communication and operational excellence include operations research, modeling and simulation, AI, big data analysis, digitally networking assets, and supply chain management. An in-depth discussion of these topics is beyond the scope of this book. The references provided at the end of the chapter can lead readers to more in-depth coverage of the topics of interest.

Companies and industries that in the past have not embraced operational excellence are now, with the help of digital tools, pursuing it. An example of operational excellence being applied in a new area not seen before, President Obama's decision to create the position of Chief Performance Officer in the initial months of his administration is a testimony to his desire to find a better way to run the Federal enterprise. That move hinted of the need to inculcate operational excellence into Federal process improvement goals which had not been formally organized before within the Federal system. The overriding fact is that recovery of the national economy implies the need for recovery of each and every entity in the economic system. As wide flung as the Federal system is suggests the increased dependency on digital tools to maintain timely communication.

An example of using new communication tools to efficiently communicate is from HID Global Corporation. The company had moved its world headquarters to Austin, Texas in 2013 and had completed the move and setup of the operations. The need to gain operational excellence and develop a culture of continuous improvement was acute. One effort the management team undertook was to do benchmarking on the Lean concept of daily management. They implemented a process that not only increased communication, but made it real time, and improved on time, quality, lead times, and inventory. They reviewed what decisions needed to be made and created a hierarchy of decision making. Pushing the decision making as far down the organization as possible would speed up response times and contribute to the culture of continuous improvement that HID was trying to instill in the new location.

Daily management meetings were set up, starting with each production department at 7 a.m. The key to the meetings, which were only 10 minutes, was to have all functions present that could help solve issues immediately on the floor—the quality engineers, product engineers, and maintenance technicians for the product line, supply chain for the product line, and customer service. The meetings were held at a communications board out on the production floor—a very large monitor that showed productivity, on time, quality, and issues in real time. The leadership of this meeting was rotated through all personnel within the production department, ensuring all operators were knowledgeable about performance and issues. Another person was timekeeper to keep the meetings to 10 minutes. Decisions that could be made by those present were assigned and noted, with due dates, on a whiteboard, and entered into the system later, but before 9 a.m. Decisions that had to be escalated to the value stream management level or above were also noted. Previous assignments were reported and if late, so noted with new due dates. Detailed problem-solving discussions were held off-line, as needed. The whiteboard and monitor were on the production floor where the operators could see it, and management would visit the boards and monitor when walking around the production floor during the day. They would ask questions of random operators to talk about problems noted.

Starting at 9 a.m., next level daily management meetings were held. These were product value stream oriented. Again, all functions were present, and the rolled up performance was shown on a large monitor. Discussions of issues and previous actions were held. Escalated issues would get assigned with due dates, and issues already assigned would be reported on. Any issues that needed escalation to upper management were also noted. Again, the meetings were kept to 10 minutes.

At 11 a.m., a Production Management daily management meeting was held in a common area on the production floor. The Senior Director led the meeting, and the head of all functions were present.

This focus on the Key Performance Indicators (KPI) engaged every employee in the understanding of performance and created a continuous improvement culture. In the common production area, there was also a white-board with current continuous improvement efforts listed so that everyone knew what efforts were being made throughout the facility. The result of this focus and effort resulted in the whole facility attaining 97% plus on time, reducing lead times by 60%, and reducing quality defects by half. This laser-sharp communication in real time was possible by the digital tools and a production system that tracked orders, production, quality, and inventory in real time, and the tools that allowed the display of the information where it was needed—the production floor.

DIGITAL COOPERATION AND PARTNERSHIPS

A natural progression in operational excellence is to leverage the capabilities of other organizations, rather than develop the capability in-house, through strategic alliances. Strategic alliance is defined as a formal alliance or "joining of forces" between two or more independent organizations for the purpose of meeting mutual business goals. Each partner in the alliance has something to bring to the "table," such as products, supply chain, distribution network, manufacturing capability, funding, capital equipment, operational expertise, know-how, or intellectual property. Strategic partnering represents cooperation whereby project management synergy ensures that each partner derives benefits beyond normal independent operation.

While there are pros and cons to partnering, the advantages often outweigh the downsides. Advantages of strategic partnering include the following:

1. It allows each partner to concentrate on operations that best match its capabilities.
2. It permits partners to learn from one another and develop competencies that may be readily utilized elsewhere.
3. It facilitates synergy that increases the outputs of both partners' resources and competencies.

Today's world requires better utilization of limited resources. More and more, cooperative partnerships are needed to achieve the highest effective use of these resources. Cooperation is a basic requirement for resource interaction and integration in any partnership. Digital communication is immediate and goes globally, sending the exact same information to all involved, a key requirement for the best cooperation. More projects fail due to a lack of cooperation and commitment than any other project factor. This lack often happens when some of the needed people are not informed in time, or get different communications. To secure and retain the cooperation of partners, the most positive aspects of a proposed partnership should be the first items of communication. Such structural communication can pave the way for acceptance of the proposal and subsequent cooperation. Then the other aspects such as profit sharing and finances, staffing, facility use, performance reporting, conflict resolution can be negotiated.

There are different types of cooperation, as summarized below:

- **Functional cooperation:** This is cooperation induced by the nature of the functional relationship between two partners. The two partners may be required to perform related functions that can only be accomplished through mutual cooperation.
- **Socially responsible cooperation:** This is the type of cooperation effected by a socially responsible relationship between two partners. This is particularly common for activities that may impact the environment. The socially responsible relationship motivates cooperation that may be useful in executing extrapreneurial partnership.
- **Regulatory cooperation:** This is usually cooperation that is based on regulatory requirements. It is often imposed through some legal authority and expectations. In this case, the participants may have no choice other than to cooperate.

- **Industry cooperation:** This is cooperation that is fueled by the need to comply with industry standards and build a consensus to advance the overall industry in which partners find themselves. For example, in the early days of cell phones, the Cellular Trade Industry Association (CTIA) developed market alliances to refute the early fears about the health impacts of using cell phones. The group sponsored several research studies at universities to confirm that cell phones were safe. Such a monumental undertaking could not have happened without market cooperation. However, caution must be exercised to ensure that cooperation does not degenerate into market collusion, which is illegal.
- **Market cooperation:** In order for each player in the market to thrive, the overall market must be vibrant. For this reason, market cooperation involves partnering of market players to advance market vitality. This usually happens in evolving markets. While similar to industry cooperation, this is focused in specific market areas. For example, regions will set up local market associations to protect certain foods so that only that region can produce the food. These associations cooperate on geographical or origin designation or manufacturing process to protect the local market against the wider food industry.
- **Administrative cooperation:** This is cooperation brought on by administrative requirements that make it imperative that two partners work together toward a common goal, such as market growth. In fact, market cooperation and administrative cooperation can coexist across organizations. One good example of administrative cooperation is the monthly meeting partnering between two professional associations in the same local community. The authors have participated in co-hosted meetings between IISE (Institute of Industrial and Systems Engineers) and ASQ (American Society for Quality).
- **Associative cooperation:** This is a type of cooperation that may be induced by collegiality. The level of cooperation is determined by the prevailing association that exists between the partners. Industry associations often cooperate under this approach.
- **Proximity cooperation:** This type of cooperation may be viewed as "Silicon-valley orientation" whereby organizations located within the same geographical setting form cooperating alliances to pursue mutual market interests. In an ideal case, being close geographically should make it possible for partners to work together. In cases where the ideal expectation does not materialize, explicit efforts must be made to encourage cooperation.

- **Dependency cooperation:** This is cooperation caused by the fact that one partner depends on another partner for some important aspect of its operation and business survival. Such dependency is usually of a mutual two-way structure. Each partner depends on the other partner for different things.
- **Imposed cooperation:** In this type of cooperation, external forces are used to induce cooperation between partners. This is often the case with legally binding requirements.
- **Natural cooperation:** This is applicable for cases where the two partners have no way out of cooperating. Physical survival requirements often dictate this type of cooperation.
- **Lateral cooperation:** Lateral cooperation involves cooperation with peers and immediate contemporaries in the marketplace. Lateral cooperation is often possible because lateral relationships create an environment that is conducive for mutual exchanges of information and operational practices. An example is the recent bailout pursuit by the big three of the US auto industry.
- **Vertical cooperation:** Vertical or hierarchical cooperation refers to cooperation that is implied by the hierarchical structure of the market in which the partners operate. For example, subsidiaries are expected to cooperate with their vertical parent organizations.

Whichever type of cooperation is available or needed in a situation, the cooperative forces should be channeled toward achieving mutual goals in the most effective way. Documentation of the level of cooperation needed will clarify roles, responsibilities, and boundaries, and win further support and sustain joint pursuits. Clarification of organizational priorities will facilitate personnel cooperation. Relative priorities of multiple partners should be specified so that a venture that is of high priority to one segment of the partnership engagement is also of high priority to all partners within the endeavor.

BLOCKCHAIN TECHNOLOGY FOR OPERATIONAL EXCELLENCE

A digital tool, the record decentralization approach of blockchain technology, originally developed for cryptocurrency purposes is rapidly making its way into the popular culture of organizational operations. Blockchain technology is defined basically as a decentralized and distributed ledger that documents and traces the origin of digital assets. Simply stated, a blockchain is a chain of blocks that contain digital information. The data that is stored inside a block depends on the type of blockchain. For example, a bitcoin block contains information about the sender, receiver, and number of bitcoins to be transferred. The first block in the chain is called the genesis block. Organizations are now adopting this blocking approach to organize and track their digital flows of information. This improves traceability and accountability and speeds communication.

One reason that blockchain has become popular is extrapreneurship. Extrapreneurship sets up a physical presence of one organization within the operation of another competing but cooperating organization. It should be emphasized that extrapreneurship is not conventional outsourcing. It is a direct setup of operations within the facility of the cooperating external organization. For example, it allows a producer to have a presence (or even a product) in an area for which it does not have internal core operational infrastructure. It is only through digital means that all such activities can be traced and accounted for. Blockchain further improves this tracking.

BITCOINING FOR OPERATIONAL EXCELLENCE

In much the same way that physical currency is used in global trade, cryptocurrency provides a digital-based movement of financial assets. More specifically for operational transactions, bitcoin is a cryptocurrency invented in 2008 by an unknown person or group of people using the name Satoshi Nakamoto. The currency began use in 2009 when its implementation was released as an open-source software. Bitcoin is a decentralized digital currency, without a central bank or a single administrator, that can be sent from user to user on the peer-to-peer bitcoin network without the need for intermediaries. As explained under blockchain above, transactions are verified by network nodes through cryptography and recorded in a public distributed ledger called a blockchain. Operational transactions often involve the transfer of assets between collaborating entities. In the digital era, entities can be physical or virtual. This makes the concept of bitcoin pertinent for operational excellence. If the movements of bitcoin can be leveraged to enhance operational interfaces of entities, then a better pathway exists for operational excellence. More and more examples of the application of bitcoining are emerging in the literature. Below are some reported or potential examples of the applications of bitcoins.

- With its global growth and acceptance, bitcoin is being used to buy goods and services. More organizations are accepting bitcoin payments. For a time, in 2018, the State of Ohio proposed to start accepting bitcoins for tax payments. The proposal was shelved pending more solid understanding of the bitcoin industry. Organizations will need to remain vigilant as the digital technology evolves more to the level of building usage confidence
- Bitcoin transactions provide a customized level of anonymity and it is relatively difficult to trace their trail. As such, bitcoins are being used in anonymous transactions that make it attractive to gamblers
- International payments can be made easily and cheaply as bitcoins are not related to any country or subject to any government regulation
- There is the freedom in the fact that there is no oversight or need for permission from any authority for transactions
- Bitcoins provide a way to transact securely online as they use very strong cryptographic algorithms
- Users and businesses like bitcoin payments because there are no credit card or exchange fees to pay
- Bitcoins can be an investment, expecting that their value will appreciate significantly in the future. In the first quarter of 2021, one bitcoin was reported to have ballooned in value to the equivalence of $50,000. But there is still much dynamism and uncertainty in the value trend
- Bitcoins are being used to shop online as increasing numbers of vendors are allowing bitcoin transactions. Users now can make payments in bitcoins on their smartphones through bitcoin wallet apps
- Unlike credit card or bank payments, there is no need to provide personal information to complete the transactions. Thus, the requirement for providing identity verification is not applicable

On a cautionary note, Tutorial Point (2021) provides some guidelines for assessing the future potential of bitcoins. Since bitcoin is a new emerging technology, unforeseen developments can make its existence and continuation difficult. Concerning its security and future, there are numerous pending questions about bitcoins:

- How far can we trust bitcoins?
- Are they a bubble that is going to burst in the face of wider applications?

- Are they a passing phenomenon and a fad that will fizzle out over time?
- Are they going to survive and dominate other currencies in future?

As of now, bitcoins are mostly unregulated. However, this may change as governments start taking a closer look due to worries about tax collection and currency control for economic development. Governments may enact legislation to regulate bitcoin, which may hugely impact the advantages that bitcoins have over other currencies. The volatility of bitcoin prices is one issue of concern. The wild fluctuations in its index are a sign of such volatility. In recent years, bitcoin prices have risen exponentially and then dipped. Overall, bitcoins are still high in value. For the purpose of operational excellence, organizations are advised to remain open-minded and vigilant to new developments with the digital bitcoin era.

EXTRAPRENEURSHIP AS A PROCESS TOOL FOR OPERATIONAL EXCELLENCE

New and agile business models can be facilitated or enhanced by digital tools and processes. Successful business enterprises often involve linkages that can be improved by digital interfaces from remote locations. Virtual business processes have progressed rapidly in recent years. So, it makes sense to continue to explore how the digital era can spark new business engagements or advance traditional business models. In this section, we discuss the specific case of how extrapreneurship can thrive in a digital era. Extrapreneurship is an extension to entrepreneurship, intrapreneurship, and other "preneurships" often embraced by business and industry. Entrepreneurship sets up independent business pursuits that compete in the open market. Intrapreneurship sets up an internal semi-autonomous business unit in an organization that may operate under a different business model from the parent company. Extrapreneurship is the next logical step in operational excellence—it sets up operating units of one competitor inside the facility(s) of another for mutual benefit. One competitor has the capabilities needed by another—the best use of resources is to set up an extrapreneurship rather than develop the capability. An extrapreneurship approach requires that the collective success be achieved through the use of a systems-oriented synergistic approach to achieve operational excellence. Having remote operations requires the accurate and timely transmission of information back to the parent company. Only the allowed information can be shared between the companies. Recognition of costs and profits to each company needs accurate tracking. All of these needs are met much more accurately and timely with today's digital tools. Without today's digital tools, the effort to operate in another's facility is reliant on manual efforts to operate and difficult. Therefore, not many competitors tried it in the past. Now the tools are there to ensure success and we should see an increase in this type of partnership in the future.

Competition can, indeed, be used as a mechanism for extrapreneurial cooperation. This happens in an environment where constructive competition paves the way for cooperation. This is sometimes designated as "coopetition" whereby the pseudo-competing organizations form a partnership to advance their mutual interests. To ensure success, communication must be clear and concise to ensure operational excellence.

The co-location strategy of extrapreneurship is what facilitates coordination and implementation of the ideals of cooperating. It represents partnership outreach of project-to-project collaboration that borders on a subsidiary relationship. Organizations must leverage existing structures of entrepreneurship and intrapreneurship to create sustainable extrapreneurship relationships. Communication facilitates cooperation while cooperation makes operational coordination possible. With a digitally framed strategic project management approach, we can build constructive extrapreneurial relationships rather than adversarial relationships when carrying out operational improvement projects. Some communication guidelines for securing extrapreneurial cooperation are as follows:

- Establish achievable goals for the extrapreneurial initiative
- Clearly outline individual commitments that are required
- Formally integrate extrapreneurship priorities with existing organizational priorities
- Anticipate and mitigate potential sources of conflict and provide a conflict resolution framework
- Alleviate skepticism by documenting the merits of the extrapreneurial initiative

Although elements of what extrapreneurship entails are being practiced by many organizations (e.g., customer site co-location), there has not been a formal or unified formulation as a formal business initiative. It may be the next best thing to direct acquisition or merger, as we have been seeing in the banking industry in the wake of the financial collapse of major financial institutions, but operational excellence—the well-executed strategic plan between companies must be well communicated and must utilize project planning for an extrapreneurship effort to succeed. That creates mutual organizational pursuits, facilitated by digital interfaces.

To kill off a competitor does not necessarily mean that the killer survives in the industry. If the industry fails, each and every entity in the industry is doomed. Cases in point include the US fabrics and steel industries, which have been lost in large numbers to global production outlets. Many other manufacturing segments are at risk of annihilation if innovative business survival strategies, which include operational excellence and the systems approach, are not developed, embraced, and pragmatically implemented. The ripple effect of the depressed world economy is spreading throughout every corner of the manufacturing sector. It behooves all manufacturers to embrace elements of extrapreneurship to ensure survival of the system as a whole.

Extrapreneurship partnering strategy implementation, just as with any business partnership, is of four different types:

1. **Joint venture:** This is a strategic alliance in which two or more partners create a legally independent company to share some of their resources and capabilities to develop a mutual competitive advantage in the marketplace.
2. **Equity strategic partnership:** This is an alliance in which two or more partners own different percentages of the company they have formed by combining some of their resources and capabilities to create a mutual competitive advantage in the marketplace.
3. **Non-equity strategic partnership:** This is an alliance in which two or more partners develop a contractual relationship to share some of their unique resources and capabilities to create mutual competitive advantage.
4. **Global reach in partnership:** This is a working partnership between companies across national boundaries and/or across industries. This is often formed between a company and a foreign government.

The key characteristic requirement for extrapreneurship is the physical co-location of units of the partnering organizations. Extrapreneurship partnering can be executed in a stepwise formation as described below.

Step 1: Strategy Development

This involves a feasibility study of the proposed alliance with respect to objectives, rationale, people, technology, process, resource base, management, challenges, and conflict resolution strategies. Strategy development requires aligning the objectives of the alliance with the overall corporate vision and mission. A key part of strategy development is an internal SWOT analysis to document strengths, weaknesses, opportunities, and threats.

Step 2: Partner Assessment

This involves assessing a potential partner's capabilities, performing benchmarking analysis, developing criteria for partner selection, developing tactics to accommodate a partner's management style, and understanding mutual resource requirements.

Step 3: Agreement Negotiation

This involves determining prospective partners' relative objectives, contributions, rewards, information protection, team composition, performance assessment measurements and procedures, policies, operating procedures, and termination clauses.

Step 4: Strategy Implementation

This involves actual operation of the alliance between the partners. It requires management's commitment to resource allocation, linking of budgets, coordination of priorities, schedule development, tracking, control, and reporting. This is an area where project management techniques are most useful. Many alliances, written so beautifully on paper, often falter due to lax implementation strategies or lack of ongoing commitment. Fast, concise, and complete communication, using digital tools, is key in executing the strategy successfully.

Step 5: Termination

All good things eventually come to an end. Partners must be realistic about this fact-based cliché. While sustainability is of utmost importance in alliance formation, what needs to be terminated must be terminated when the time comes. Force-feeding an alliance just for the sake of sustaining a relationship could only lead to a waste of time, squandering of resources, and lost opportunities. It is time to terminate an alliance when its objectives have been met or cannot be met, or when a partner adjusts priorities or diverts resources toward other initiatives.

LEVERAGING PROJECT MANAGEMENT FOR EXTRAPRENEURSHIP

Project management is the process of managing, allocating, and timing resources to achieve a given goal in an efficient and expeditious manner. Project management is required now more than ever for getting things done and moving concepts to reality. In the past, an organization's reach was often local, the projects smaller. They could be managed with simpler tools. Today, however, with partnerships, global supply chains and the demand for fast, reliable solutions, simple tools are not enough. Project management is a systematic approach to developing and communicating the implementation plan. The concept of extrapreneurship uses project management strategies as the basis for practical accomplishment of what cooperation sets out to achieve and to communicate to distant locations. Without a project plan, the people who have to make the concept reality cannot execute the concept, communicate progress, or identify issues that arise in a timely manner. A solid project plan, using digital tools to communicate swiftly, is the foundation to operational excellence. For example, as the world economy recovers from the current pandemic, formal application of a systems approach with strong project management and communication creates an effective audit process that enhances overall operations through the following:

- Business and industry leadership benchmarking and development
- Involvement of experts across organizational boundaries
- Diversification of business and operating processes
- Systems view of the national economy

OPERATIONAL EXCELLENCE IN NONTRADITIONAL AREAS

We often think of operational excellence as the focus of business and industry alone. But there are nontraditional areas where operational excellence is also a desirable pursuit. One such area is the military. Specifically, in this section, we discuss the case of operational excellence in the US Air Force as presented by a published article from the Air Force Institute of Technology (AFIT). In recent years, the US Air Force has announced several long-term efficiency initiatives, many of which center on organizational transformation programs. For these initiatives to be fully successful, the Air Force must leverage the capabilities of the IEs among its ranks. The current process for managing the utilization of scientists and engineers in the Air Force does not adequately differentiate among technical disciplines and specialties between and within career fields. Moreover, the system lacks stability in that it changes rapidly and is rarely updated. Badiru and Thomas (2013) present an example of enterprise transformation in the specific case of the US Air Force workforce development through advanced education. The objective is an appropriate education and utilization of the workforce. The premise of the example is that the Air Force can effectively achieve enterprise transformation through better operational efficiency using industrial engineering techniques. The Air Force has a large number of degreed IEs, but they are rarely placed in assignments that can directly utilize the skills of the profession, particularly as it relates to the digital era of operational opportunities. Using a systems view to harness industrial engineering skills and tools throughout the Air Force will lead to a more widespread transformation of operating practices and procedures to achieve operational excellence.

Any attempt to achieve organizational transformation must be based on an educational foundation for all employees at the respective levels of responsibilities and duties throughout the organization. IE, by virtue of its legacy of efficiency, provides a strategic option for achieving the desired organizational transformation. As discussed in Chapter 1, IE emerged out of industry's need for efficient work systems and processes for better utilizing workers in factory and operations in 1900. The key elements of IE practice are people, processes, and products with a focus on optimum performance and continuous improvement, often in allocating scarce resources. Enterprise transformation is the coordinated pursuit of operational improvements by way of an integrated process of the application of various tools and approaches for improving operations while reducing resource expenditures. Applying these concepts throughout the military will lead the military to operational excellence.

Lean is a very old concept originally introduced for improving manual work in factory and construction operations. The modern Lean principles that have evolved from the early IE practitioners were developed for production and manufacturing systems. The applications of Lean principles for services came much later in the evolution of Lean in the United States. It is of historical note that Japan was applying Lean to service areas for a very long time. They didn't let the non-Japanese visitors touring Toyota and other manufacturing facilities see that they were applying it in all areas of the companies, and in nonmanufacturing companies, until the 1990s.

Military needs are predominately for service operations, rather than production operations. The military enterprise substantively and directly affects the national economy either through direct employment, subcontracts, military construction, or technology transfer. Thus, it is fitting to expect that military enterprise transformation can have direct impacts on general civilian enterprise transformation programs. Kotnour (2011) presents the fundamental elements and challenges of enterprise transformation with a view toward developing a universal framework for assessing the effectiveness of transformations. Some of the key elements he suggested are as follows:

- Successful change is leadership driven
- Successful change is strategy driven
- Successful change is project managed
- Successful change involves continuous learning
- Successful change involves a systematic change process

The above elements are all within the scope of the application of industrial engineering in the pursuit of operational excellence. What the US Air Force needs to achieve organizational transformation is already available organically within the Air Force. That is, the availability of IEs. Susan Blake (2011), who works at Tinker Air Force Base, puts it aptly with the definition below:

> IEs make systems function better together with less waste, better quality, and fewer resources.

As it is with every organization, the major goal of the US Air Force's long-term efficiency initiatives is to eliminate waste, achieve quality improvement in products and services, and optimize resource utilization. In other words, achieve operational excellence. The US military, particularly the Air Force, has a cadre of IEs in the development engineer (or field engineer) career field, but they have not been formally engaged nor identified as a resource for implementation of enterprise transformation. As a specific example, the initiation and implementation of AFSO21 (Air Force Smart Operations for the 21st Century) gained considerable success in elevating awareness that is essential for altering the culture of Air Force operations toward Lean and efficient processes (Badiru, 2007, 2014a, 2014b, 2019a, 2019b). Numerous successes have been achieved. Still, there is a long journey ahead for a complete implementation and sustainment of the program. Although the case of the US Air Force is cited for discussion purposes, the transformation strategies are equally applicable to other service branches in the US military. The application of industrial engineering from a systems perspective can directly contribute to the priorities established by the Air Force, as summarized below:

1. Reinvigorate the Air Force Nuclear Enterprise
    a. Focus on precision and reliability in Air Force processes
    b. Fulfill mandate for accountability

2. Partner with the Joint and Coalition team to win today's fight
    a. Aggressively adapt Air Force ways and means across the spectrum of Command and Control (C2) and Intelligence, Surveillance, and Reconnaissance (ISR)
    b. Focus on joint capabilities, interoperability, and mutual operational trust
3. Develop and care for Airmen and their families
    a. Train personnel to be ready for 21st Century challenges
    b. Reinforce Warfighting Ethos and expeditionary combat mindset
    c. Accommodate demands on families
4. Modernize aging air and space inventories
    a. Get all services to "reset" and build for the future
    b. Build a balanced force for the future
5. Achieve Acquisition Excellence
    a. Focus on process, people, and performance

In the prevailing digital era, the priorities outlined above will need to be considered on a bedrock of digital engineering, as discussed earlier in this chapter. All of the priorities can only be achieved through a systems view of overall operations. A systems approach facilitates a comprehensive consideration of all the facets of an organization (Badiru, 2006). Drew (2008) presents a collection of essays postulating the need for recapitalizing the Air Force intellect through military education. One argument that he presents is that problems articulated over three decades ago still remain unsolved and relevant to the present-day operational challenges for the Air Force. If these problems have remained unsolved for so long, we believe the time has come to apply industrial engineering techniques (Badiru and Thomas, 2009) to achieve sustainable solutions. Garrett and Rendon (2007) present program management strategies for the US military. Their treatise covers operational issues related to the following:

- Requirements management
- Program leadership and teamwork
- Risk and financial management
- Supply chain management and logistics
- Contract management and procurement

Sustainability of operational excellence in the military depends on several factors, including need for pre-analysis, operational requirements, work specifications, organizational capabilities, service infrastructure, administrative processes, and process standardization. The interplay of all these factors requires an integrated systems approach to operational excellence, leveraging the new digital era.

VALUE OF VERSATILITY AND OPERATIONAL TRANSFORMATION IN THE MILITARY

Although IE techniques are not applied for officer assignment processes, the techniques are used in the pursuit of program effectiveness and operational efficiencies. It can be seen from the definition that a systems orientation permeates the work of IEs. This is particularly applicable to the military because military operations and functions are designed to encompass the linking of subsystems. Acosta et al. (2010) discuss operational improvement in the context of a global engineering framework. They present strategic issues related to improving communication and decision processes in a global operating environment, which is typical of today's US military expeditionary missions, just as it is in the civilian sector. In order to achieve the desired improvement, sweeping organizational changes are essential as detailed by Robbins and Judge (2010). They address issues of motivation concepts and personnel necessities following the path of Maslow's hierarchy of needs. It is when the needs of personnel are met at each stage of an organization that the organizational behavior can be altered positively to achieve collective goals and objectives.

Military operational requirements are changing rapidly in response to the fast changes around the world. A big part of this challenge is the shrinking resource base. In Lean times, military organizations must develop strategies to operate more efficiently and effectively. Even in times of plenty, higher efficiency and effectiveness are required because Lean times are always on the horizon. Continuous improvement is an essential component of AFSO21, which is primarily composed of Lean principles. Lean, in its simplest form, implies the elimination of waste. If waste can be reduced in operations, opportunities for continuous improvement can be maximized. Many of the current Air Force processes contain inherent waste, lack of consistency, and misallocation of intellectual assets.

Operational excellence requires getting the most out of every asset. The right asset (human, technology, and facilities) must be matched against the right requirements. A strategy for accomplishing this is to take inventory of available assets and use multidimensional matrix allocation techniques to match assets to requirements. A good example of this is the large inventory of IEs within the Air Force that are not being effectively utilized in ways to leverage their training and expertise for operational improvement.

LEAN PRINCIPLES FOR OPERATIONAL EXCELLENCE

Lean, or continuous process improvement, is an ongoing systematic effort to improve day-to-day operations to remain productive and operationally efficient. For example, AFSO21 is a coordinated pursuit of operational improvements throughout the Air Force by way of an integrated process of applications of various tools and approaches for improving operations while reducing resource expenditures. Productivity, quality of service, process enhancement, flexibility, adaptability, work design, schedule optimization, and cost containment are within the scope of AFSO21. Increases in operational efficiency are best accomplished through gradual and consistent closing of gaps rather than the pursuit of one giant improvement step. The practice of drastic or sudden improvement often impedes process optimization goals.

Lean is a way of doing business, a culture, that focuses the entire organization on the identification and elimination of sources of waste in operations. By comparison, the Six Sigma program uses a subset of Lean tools to identify and eliminate the sources of defects through the reduction of variability, a type of waste. When Lean and Six Sigma are combined, an organization can reduce both waste and defects in operations. Consequently, the organization can achieve higher performance, better employee morale, more satisfied constituents, and more effective utilization of limited resources.

The basic principle of Lean is to take a close look at the elemental composition of a process to eliminate non-value-adding elements or waste. Lean and Six Sigma techniques use analytical and statistical techniques as the basis for pursuing improvement objectives. But the achievement of those goals depends on having a structured approach to the activities associated with what needs to be done. If an IE approach is embraced at the outset, it will pave the way for achieving Six Sigma results and make it possible to realize Lean outcomes. The key in any operational management endeavor is to have a structured plan so that diagnostic and corrective steps can be pursued.

If inefficiency is allowed to creep into operations, it would take much more time, effort, and cost to achieve a Lean Six Sigma cleanup. To put the above concepts in a military perspective, Six Sigma implies executing C2 processes such that errors are minimized in the long run. Likewise, the techniques of Lean ensure that only value-adding C2 actions are undertaken. This means the elimination of waste. This brings to mind **Parkinson's law** of bureaucracy, which states that "work expands to fill the time available"; as a result of which unnecessary activities are performed. Military leaders must ensure that functions do not extend needlessly just to use up available time and resources. Short and effective functions are better than protracted ones that result in counterproductive results. At this point, it is of note to mention that the popular **Murphy's law**, which states that "whatever can go wrong will go wrong," originated in 1947 from a graduate of the AFIT, **Captain Edward A. Murphy**. After retiring from active military service, Edward A. Murphy worked as a civilian engineer, who spent a lifetime studying reliability and safety in order to prevent human error. He coined "Murphy's Law" with an offhand remark while working at Edwards Air Force Base. While stationed at Wright-Patterson Air Force Base in the late 1940s, Edward Murphy was a member of the Engineers Club of Dayton, of which the first author of the book is also a member. The cautionary note embodied in Murphy's law can pave the way for operational excellence. As we often experience in the virtual environment created by the COVID-19 pandemic, whatever can go wrong in the digital virtual environment has been known to go wrong. For operational excellence, careful planning, precautions, and contingency actions can help preempt operational failures in the digital environment. Below is a summary of the common rules, laws, and principles encountered in an operational setting:

Parkinson's law: Work expands to fill the available time or space.
Peter's principle: People rise to the level of their incompetence.
Murphy's law: Whatever can go wrong will go wrong.
Badiru's rule: The grass is always greener where you most need it to be dead.

EFFECTIVE WORKFORCE UTILIZATION

An assignment to a task implies being involved in various aspects of the task's requirements. Communication complexity is one aspect that is often overlooked in the assignment of personnel to tasks where they do not add any value. The increased requirement for multi-person and hierarchical communication is the subtlest of wastes in any organization because it not only impedes overall organizational performance, but it also reduces the utilization of critical resources. A structural review of communication, cooperation, and coordination of personnel can identify where resource utilization improvement can be achieved. Often the assignment requirements within the Air Force are not in sync with the prevailing constraints. The resolution of such problems requires integrative communication, cooperation, and coordination. This is achieved through a systems-based approach. Basic questions of what, who, why, how, where, and when should always be addressed strategically when doing personnel and resource assignments. It highlights what must be done and when, with respect to the following:

- Does each participant know what the objective is?
- Does each participant know his or her role in achieving the objective?
- What obstacles may prevent a participant from playing his or her role effectively?

IMPACT OF COMMUNICATION COMPLEXITY

Communication complexity increases with an increase in the number of decision points. It is one thing to wish to communicate broadly, but it is another thing to create complexity when more decision points are involved. The Air Force structure typically involves multilayer decision processes. While this provides for more chains of command with embedded checks and balances, it can impede overall systems efficiency by creating unnecessary layers of decision. The versatility, ease, and flexibility of the digital era may lure us into a situation of complex communication channels that impact complexity on operations to the point of diminishing value rather than improving performance. As the number of those involved increases, so increases the complexity of the operations.

The work environment must be designed to facilitate cooperation in support of enterprise transformation programs. Team cooperation is influenced by proximity, functional relationships, professional affiliations, social relationships, official capacity influence, formal authority influence, hierarchical relationships, lateral camaraderie, intimidation, and enticement. Operational coordination can be achieved through teaming, delegation, empathetic supervision, partnership, token-passing, and deliberate operational baton handoff. These are all systems-oriented processes that should be leveraged for cooperative enterprise transformation for the purpose of facilitating operational excellence. In summary, communication complexity can impede the overall performance of an organization. In the analogy to computational science, communication complexity studies the amount of communication required to solve a problem when the input to the problem is distributed among two or more parties. Communication complexity can be approximated as the number of potential unique conversations that can occur in a group. Below are some factors that could affect communication complexity:

- Interpersonal conflicts
- Office politics
- Leadership style
- Communication tools
- Plainness of the communication content
- Analogy versus digital communication
- Traceability of communication contents
- Message framing and context

EMERGENCE OF 5-G TECHNOLOGY FOR OPERATIONAL EXCELLENCE

Using digital communication tools can help reduce complexity, conflicts, and politics, enhance leadership, and speed up communications. Choosing the right tool, and designing the use of the tool is key to successful communication. In this regard, the emerging focus on 5-G technology should be embraced for operational excellence. 5G is the fifth generation of mobile technology, which is an advancement beyond the present 4G technology. 5G provides the platform for new and emerging technologies, such as Internet of things (IoT), AI, and big data, to improve the way we live, work, and exercise operational excellence.

REFERENCES

Acosta, Carlos, V. Jorge Leon, Charles Conrad and Cesar O. Malave (2010), *Global Engineering: Design, Decision Making, and Communication*, Taylor & Francis Group/CRC Press, Boca Raton, FL.

Badiru, Adedeji B. (2006), editor, *Handbook of Industrial & Systems Engineering*, Taylor & Francis Group/CRC Press, Boca Raton, FL.

Badiru, Adedeji B. (2007), "Air Force Smart Ops for the 21st Century," *OR/MS Today*, Vol. 34, No. 1, February, p. 28.

Badiru, Adedeji B. (2014a), "Quality Insights: The DEJI® Model for Quality Design, Evaluation, Justification, and Integration," *International Journal of Quality Engineering and Technology*, Vol. 4, No. 4, pp. 369–378.

Badiru, Adedeji B. (2014b), editor, *Handbook of Industrial & Systems Engineering*, 2nd edition, Taylor & Francis Group/CRC Press, Boca Raton, FL.

Badiru, Adedeji B. (2019a), *Systems Engineering Models: Theory, Methods, and Applications*, Taylor & Francis Group/CRC Press, Boca Raton, FL.

Badiru, Adedeji B. (2019b), *The Story of Industrial Engineering: The Rise from Shop-Floor Management to Modern Digital Engineering*, Taylor & Francis Group/CRC Press, Boca Raton, FL.

Badiru, Adedeji B. and Marlin U. Thomas (2009), editors, *Handbook of Military Industrial Engineering*, Taylor & Francis Group/CRC Press, Boca Raton, FL.

Badiru, Adedeji B. and Marlin U. Thomas (2013), "Quantification of the PICK Chart for Process Improvement Decisions," *Journal of Enterprise Transformation*, Vol. 3, No. 1, pp. 1–15.

Blake, Susan (2011), "5-Minute Explanation of What an IE Does," *Industrial Engineering*, Vol. 43, No. 10, October, p. 12.

Drew, Dennis M. (2008), *Recapitalizing the Air Force Intellect: Essays on War, Airpower, and Military Education*, Air University Press, Maxwell AFB, Alabama.

Garrett, Gregory A. and Rene G. Rendon (2007), *U.S. Military Program Management: Lessons Learned and Best Practices*, Management Concepts, Vienna, VA.

Gass, Saul (2002), "Great Moments in the History of Operations Research," *OR/MS Today*, Vol. 29, No. 4, October, pp. 29, 31–37.

Kotnour, Tim (2011), "An Emerging Theory of Enterprise Transformations," *Journal of Enterprise Transformation*, Vol. 1, No. 1, pp. 48–70.

Robbins, Stephen P. and Timothy A. Judge (2010), *Essentials of Organizational Behavior*, 10th edition, Prentice-Hall, Upper Saddle River, NJ.

Tutorial Point (2021), "Bitcoin—Future," https://www.tutorialspoint.com/bitcoin/bitcoin_future.htm, Accessed, March 28, 2021.

Systems Thinking in the Digital Era 3

INTRODUCTION

It is a systems world and must capitalize on what systems thinking brings to all operations. Because of its multiplicity of potentials, a systems approach is the way to go in order to leverage the full potential of the digital era. Industrial engineering (IE) is a systems-oriented profession. For this reason, it is often interfaced with systems engineering (SE). This alignment has led to the emergence of the combined field of industrial and systems engineering (ISE). Thus, there is a need to define these two professions in order to have a clear perspective about how they interrelate. Throughout this chapter, these two fields are intertwined, and the discussions that follow, based on Oke (2014), are applicable to either. Perhaps, the first classic and widely accepted definition of IE was offered by the then American Institute of Industrial Engineering (AIIE) in 1948. Others have extended the definition:

> Industrial Engineering is uniquely concerned with the analysis, design, installation, control, evaluation and improvement of socio-technical systems, in a manner that protects the integrity and health of humans, social and natural ecologies. A socio-technical system can be viewed as any organization which people, materials, information, equipment, procedures and/or energy interact in an integrated fashion, throughout the life cycles of its associated products, services, or programs.
>
> Through a global system's perspective of such organizations, industrial engineering draws upon specialized knowledge and skills in the mathematical, physical, and social sciences, together with the principles and methods of engineering analysis and design to specify product, and evaluate the results obtained from such systems, thereby assuring such objectives as performance, reliability, maintainability, schedule adherence and cost control.

DOI: 10.1201/9781003052036-3

By whatever name it is called, the value of industrial engineering is measured not by its words, but by its deeds. Case examples of industrial engineering deeds are prevalent in all kinds of business, industry, and government. Such deeds are facilitated by the use of systems thinking.

According to the International Council on Systems Engineering (INCOSE), SE is an interdisciplinary approach and means to enable the realization of successful systems. Such systems can be diverse, encompassing people and organization; software and data; equipment and hardware; facilities and materials; and services and techniques. The system's components are interrelated and use organized interaction toward a common purpose. From the viewpoint of INCOSE, SE focuses on defining customer needs and required functionality early in the development cycle, documenting requirements, and then proceeding with design synthesis and systems validation while considering the complete problem. The philosophy of SE teaches that attention should be focused on what the entities do before determining what the entities are. A good example to illustrate this point may be drawn from the transportation system. In solving a problem in this area, instead of beginning the problem-solving process by thinking of a bridge and how it will be designed, the systems engineer is trained to conceptualize the need to cross a body of water with certain cargo in a certain way.

The systems engineer then looks at bridge design from the type of bridge to be built. For example, is it going to be suspension or superstructure design? From this stage, he would work down to the design detail level where civil engineers get involved, considering foundation soil mechanics and placements of structures. The contemporary business is characterized by several challenges. This requires the industrial engineers and systems engineers to have similar or complementary skills, knowledge, and technical know-how in the collection, analysis, and interpretation of data relevant to problems that arise in the workplace to place the organization well above the competition. There are interactions between the various and diverse general areas of IE and SE. The diverse areas include sociology, economics, mathematics, psychology, accounting, and electronic computing. Each of these areas specifically points out some positive contributions to the growth of IE and SE, aided further by the digital era. Thus, ISE is a combination of the two disciplines.

The key to applying a system view in the digital era is to have an appreciation for the inputs, process, and outputs related to scientific management of efficiency, effectiveness, and productivity. This is aptly framed in the context of the elements presented in Figure 3.1. Inputs relate to how the digital era has changed and impacted the inputs to an operational system. Process relates to how the digital era has changed and impacted the operating processes (policies and procedures) embodied in the system. Outputs convey how the digital era has changed and impacted the outputs coming out of the system. In all

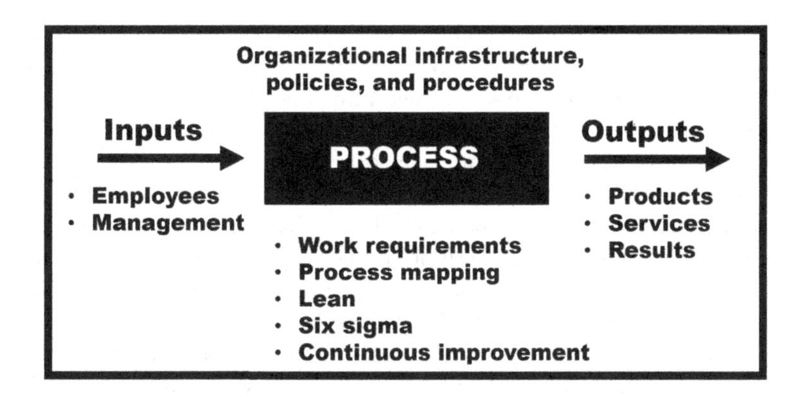

FIGURE 3.1 Input, process, and output relationships for operational excellence.

these cases, the change is most noticeable in how data is collected, stored, retrieved, manipulated, transmitted, and utilized. IE tools and techniques are directly applicable and leveraged at every stage of the move toward operational excellence.

SYSTEMS MODELS FOR OPERATIONAL EXCELLENCE

Several SE models are available in practice (Badiru, 2019). The models facilitate practical execution of operations in the digital environment. In many cases, the applications are customized for internal organizational applications and are not fully documented in the open literature. The most common models include the waterfall model, the V-model, the spiral model, the walking skeleton model, DEJI Systems Model, and others. Many of them originated in the software development industry. Selected ones are described below.

The Waterfall Model

The waterfall model, also known as the linear-sequential life cycle model, breaks down the SE development process into linear-sequential phases that do not overlap one another. The model can be viewed as a flow-down approach to engineering development. The waterfall model assumes that each preceding phase must be completed before the next phase can be initiated. Additionally, each phase is reviewed at the end of its cycle to determine whether or not the project aligns with the project specifications, needs, and requirements. Although the orderly progression of tasks simplifies the development process, the waterfall model is unable to handle incomplete tasks or changes made later in the life cycle without incurring high costs. This makes sense for the waterfall model since water normally flows downward, unless forced to go upward through a pumping device, which could be an additional cost. Therefore, this model lends itself better to simple projects that are well defined and understood.

Case study: This model was instrumental in updating the Conway Regional Medical Center in Arkansas. In 2010, the regional center still did not have an electronic database and information management system for its home health care patients. Consequently, the hospital applied the waterfall method to their acquisition of software to handle their needs. First, the hospital management defined their problem as needing a way to maintain a database of documentation and records for the home health care patients. The hospital then elected to buy, rather than design, the software necessary for this project. After defining their system requirements, the hospital's administration team purchased what they evaluated to be the most suitable software option. However, the hospital performed systems testing before integrating the software into the home health care system. Upon completion of testing, the hospital found that the code needed to be updated once every six weeks. This update was factored into their operation and maintenance plan for use of this new system. The system was finally deemed a resounding success once the system was implemented, tested, and the operations and maintenance schedules were created. Through the use of the waterfall model, the software got on track for installation as the primary home health care software for the Conway Regional Medical Center.

The V-Model

The V-model, or the verification and validation model, is an enhanced version of the waterfall model that illustrates the various stages of the system life cycle. The V-model is similar to the waterfall model in that they are both linear models whereby each phase is verified before moving on to the next phase. Beginning from the left side, the V-model depicts the development actions that flow from concept to the integration and verification activities on the right side of the diagram. With this model, each phase of the life cycle has a corresponding test plan that helps identify errors early in the life cycle, minimize future issues, and verify adherence to project specifications. Thus, the V-model lends itself well to proactive defect testing and tracking. However, a drawback of the V-model is that it is rigid and offers little flexibility to adjust the scope of a project. Not only is it difficult, but it is also expensive to reiterate phases within the model. Therefore, the V-model works best for smaller tasks where the project length, scope, and specifications are well defined.

As an example, the V-model was used with great success to develop the Smart Transportation Authority by the Chattanooga Area Regional Transportation Authority (CARTA, https://www.gocarta.org/). It was one of the first smart transport systems in the United States. The V-model was used to guide the design of the new system and to integrate it into the existing system of buses, electric transport, and light rail cars. The new smart system introduced a set of features such as customer data management, automated route scheduling to meet demand, automated ticket vending, automated diagnostic maintenance system, and computer-aided dispatch and tracking. These features were revolutionary for a mid-sized metropolitan area. CARTA was able to maintain their legacy transport while integrating their new system. They were able to do this by splitting the V-model into separate sections. CARTA had a dedicated team to manage each segment of the V-model. The flat portion of the V-shape represented the legacy transportation systems. Individual teams focused on the definition, test, and integration of all the new components of the overall system. Separating the two sides of the V-model (legacy and innovation) enabled CARTA to maintain valuable functionalities while adding new features to the integrated system that enhanced usability, safety, and efficiency.

Spiral Model

The spiral model is similar to the V-model in that it references many of the same phases as part of its color-coordinated slices, which indicates the project's stage of development. This model enables multiple flows through the cycle to build a better understanding of the design requirements and engineering complications. The reiterative design lends itself well to supporting popular model-based SE techniques such as rapid prototyping and quick failure methods. Additionally, the model assumes that each iteration of the spiral will produce new information that will encourage technology maturation, evaluate the project's financial situation, and justify continuity. Consequently, the lessons learned from this method provide a data point with which to improve the product. Generally, the spiral model meshes well with the defense life cycle management vision and integrates all facets of design, production, and integration.

The spiral model is the foundation of the military's RQ-4 Global Hawk Operational Management and Usage platform. The Global Hawk was phased into operation in six distinct spirals, with each spiral adding new capabilities to the airframe. The first spiral was getting the aircraft in the sky and having a support network to keep it in the air. Everything from pilots to maintenance was optimized to keep the Global Hawk in the air as much as possible. The subsequent phases added imagery (IMINT), signal (SIGINT), radar, and survivability capabilities to the airframe. Each of these capabilities was added one at a time in a spiral development cycle to ensure that each one was integrated into the airframe to the operational standard, and could be adjusted to meet this standard before moving on to the next capability. The benefit of incrementally adding capabilities in a spiral fashion greatly helped the Global Hawk stay on budget and schedule for operational rollout.

Defense Acquisitions University SE Model

The new Defense Acquisitions University (DAU) model for SE (Badiru, 2019) also originates from the V-model. However, unlike the traditional V-model, it allows for process iteration similar to the spiral model. A unique attribute of the DAU process is that its life cycle does not need to be completed in order to gain the benefit of iteration. Whereas the spiral model requires the life cycle process to be completed, the DAU model can refine and improve products at any point in its phase progression. This design is beneficial to making early-stage improvements, which helps systems engineers to avoid budgeting issues such as cost overruns. Moreover, the model allows for fluid transition between project definition (decomposition) and product completion (realization), which is useful in software production and integration. Overall, the DAU model is a fluid combination of the V-model and spiral model.

This tailored V-model was used by Air Force researchers to create a system to aid battlefield airmen in identifying friendly forces and calling in close air support with minimal risk to ground troops. The model was used to find an operational need and break it down into a hierarchy of objectives. The researchers used the hierarchy to design multiple prototypes that attempted to incorporate all of the stated objectives. They used rapid prototyping methods to produce these designs, and they were then tested operationally within battlefield airmen squadrons. Ultimately, the production of a friendly marking device was achieved and this valuable capability was able to be delivered to the warfighter.

Walking Skeleton Model

The walking skeleton model is a Lean approach to incremental development, popularly used in software design. It centers on creating a skeleton framework for what the system is going to do and look like. This basic starting point of the system will have minimal functionality, and the systems engineer will work to add muscles to the skeleton. The first step creates a system that may do a very basic yet integral part of the final system design. For example, if one were to design a car using this method, the skeleton would be an engine attached to a chassis with wheels. Once the first basic step is done, the muscles begin to be added to the skeleton. These muscles are more refined and are added one at a time, meaning that each new feature of the system must be completed before working on the next feature. Furthermore, it is highly recommended that the most difficult features of the system are the first muscles to be added. System components that take a lot of time, require contracting/outsourcing, or are the primary payload must be the first to be completed. This will become the heart of the skeleton and the rest of the architecture can be optimized to ensure the critical capability of the system is preserved and enhanced.

An example of engineering using the walking skeleton model is the Boston Dynamics walking dogs. The first thing that the engineers did for these robotic creatures was to create a power source and mobility framework. From there, the engineers were able to then go piece by piece and add more functionality to the project, such as the ability to open doors, pick up objects, and even carry heavy loads. The lessons they learned from adding muscles to their skeleton allowed them to move in leaps and bounds, and the benefits were felt across their entire network of products.

The walking skeleton technique varies with the system being developed. In case of a client-server system, it will be a single screen connected for navigating to the database and back to the screen. In a front-end application system, it acts as a connection between the platforms and compilation takes place for the simplest element of the language. In a transaction process, it is walking through a single transaction.

Following are the techniques that can be used to create a walking skeleton:

- **Methodology shaping:** Gathering information about prior experiences and using it to come up with the starter conventions. The following two steps are used in this technique:
    1. Project interviews
    2. Methodology shaping workshop
- **Reflection workshop:** A particular workshop format for reflective improvement. In the reflection workshop, team members discuss what is working fine, what improvements are required, and what unique things will be added next time.
- **Blitz planning:** Every member involved in project planning notes all the tasks on the cards, which will then be sorted, estimated, and strategized. Then the team decides on the resources such as cost, time and discuss about the roadblocks.
- **Delphi estimation:** A way to come up with a starter estimate for the total project. A group consisting of experts is formed and opinions are gathered with an aim to come up with highly accurate estimates.

- **Daily stand-ups:** A quick and efficient way to pass information around the team on a daily basis. It is a short meeting to discuss status, progress, and setbacks. The goal is to keep meetings short. This meeting is to identify the progress and roadblocks in the project.
- **Agile interaction design:** A fast version of usage-centered design, where multiple short deadlines are used to deliver working software without giving important considerations to the activities of designing. For example, to simplify the user interface test LEET (software coding slang derived from elite), a record/capture tool is used. "Leet" is derived from the word "elite," which is a code typically used by hackers in the 1980s to refer to a cult-like "leet speak language" to throw off competing software developers. Leetspeak (or leet) is an alternate representation of text that replaces letters with numbers or character combinations. The coding and decoding are not always unique, depending on the characters and numbers included in the leet template.
- **Process miniature:** A learning technique when a new process is unfamiliar and time-consuming. When the process is complex, more time is required for new team members to understand how different parts of the process fit. Time taken to understand the process is reduced with use of process miniature.
- **Side-by-side programming:** Pair programming. An alternative of pair programming is "Programming in pairs." Here two people work on one assignment by taking turns in providing input and mostly on a single workstation. It results in better productivity and the cost of fixing bugs is less.

  Programmers work without interfering in their individual assignments and review each other's work easily.
- **Burn charts:** This tool is used to estimate actual and estimated amount of work against the time.

DIGITAL OBJECT-ORIENTED ANALYSIS AND DESIGN

Digital object-oriented analysis and design (OOAD) is an agile methodology approach toward SE and eschews traditional systems design processes. Traditional methods demand complete and accurate requirement specification before development; agile methods presume that change is unavoidable and should be embraced throughout the product development cycle. This is a foreign concept to many systems engineers that follow precise documentation habits and would require an overhaul of project management architecture in order to work. If the necessary support is in place to allow for this approach, it works by grouping data, processes, and components into similar objects. These are the logical components of the system that interact with each other.

For example, customers, suppliers, contracts, and rental agreements would be grouped in a single object together. This object would then be managed by a single person with complete executive control over the data and relationships within. This approach is people based, relying on the individual competencies and exquisite knowledge of their respective object. The systems manager then needs to link all of the people and their objects together to create the final system. The approach hinges upon each person perfecting the object that they are in charge of, and the systems engineer puts all of the pieces together. It puts all of the design control in the hands of the individual engineers. The most popular venue for use of this type of SE is software engineering. It allows experts within their fields to focus on what they do best for a program. OOAD does not allow for convenient system oversight, process verification, or even schedule management, and as such makes it very difficult to get consistent project updates. While it may be conducive to small team projects, this method is unlikely to be feasible for any large-scale projects.

ISEs in the Digital Era

Oke (2014) gave a good overview of how ISEs integrate systems perspectives into operational performance. ISEs are the most needed and preferred engineering professionals today, due to their ability to understand and manage complex organizations and how to apply the systems approach. They are trained to ensure the design, development, and installation of optimal methods in coordinating people, materials, equipment, energy, and information. The integration of these resources is needed in order to create products and services in a business world that is becoming increasingly complex and globalized and demanding of faster response times.

The complexity of today's industrial-organizational environment requires the new digital tools to function effectively. Some of the factors impacting digital implementations include foreign competitors, global sources of supply, political instability around the world, international quality standards, stricter product requirements, trade restrictions, increasing workforce cost, capital-intensive business expansion, shifting customer demands, higher quality demands, more mobile workforce, environmental changes, and economic instability. Understanding that these are all part of the system being analyzed is important to the ability to improve the system. ISEs have the skills to understand and integrate the new digital tools into the systems for optimization and the best results.

Many of an organization's processes, such as wage and salary administration systems and job evaluation programs, have also been developed by ISEs using new digital tools, leading to their being absorbed into management positions. ISEs share similar goals with health and safety engineers in promoting product safety and health in the whole production process through the application of this knowledge to industrial processes and such areas as mechanical, chemical, and psychological principles. They are well grounded in the application of health and safety regulations while anticipating, recognizing, and evaluating hazardous conditions and developing hazard control techniques.

ISEs, using these enhanced tools, can assist in creating leaner, more efficient and profitable business practice that improves customer services and the quality of products. This improves the resource utilization in organizations, leading to increased competitiveness. However, ISEs must still be engaged in traditional labor or time standards setting and in the redesign of organizational structure in order to eliminate or reduce some forms of frustration and wastes in manufacturing. Paying attention to the basics is still essential for the long-term survivability and health of the business.

ISEs make work safer, easier, and more rewarding through better designs that reduce production cost and allow the introduction of new technologies. This is needed in the digital era—companies need to be competitive and get good at it faster than the competition. ISE skills are essential to this. ISEs focus on rapid response through removing waste and improving efficiency. The digital era has given ISEs improved analytical tools to improve performance, but only when the new technologies are integrated into the system. Integrating requires the systems approach—seeing the whole system while improving a portion, so that the improvement does not harm the overall system, and is what ISEs have been doing since the profession began. This ability to see the system and understand how the parts interact is crucial to rapidly changing and improving the system as a whole. This improves the lives of the populace by being able to afford to use technologically advanced goods and services. In addition, they show ways to improve the working environment, improving efficiencies and increasing cycle time and throughput, thus helping all types of organizations get their products and services out quicker. ISEs can incorporate new digital methods by which businesses can analyze their processes and make improvements on them. It is focused on optimization—doing more with less—and helps to reduce waste in the society. ISEs also provide assistance in guiding society and business to care more for their workforce while improving the bottom line.

The radical growth in global competition, constantly, and rapidly evolving corporate needs, and the dynamic changes in technology are some of the important forces shaping the world of business. Thus, stakeholders in the economy are expected to operate within a complex but ever-changing business environment. Against this backdrop, the dire need for professionals that are reliable, current, effective, and relevant becomes obvious. ISEs are certainly needed in the economy for radical change, value creation, and significant improvement in productive activities.

ISEs must be focused and have the ability to think wide in order to make a unique contribution to society. To complement this effort, the organization itself must be able to develop effective marketing strategies (aided by the powerful Internet tool) as a competitive advantage such that the organization can position itself as the best in the industry.

GLOBAL SYSTEMS CHALLENGES

The challenges facing the ISE may be viewed from two dimensions: Those faced by ISEs in developing and underdeveloped countries and those faced by ISEs in the developed countries. For the developed countries, there is a high level of technological sophistication that promotes and enhances the professional skills of the ISE.

Unfortunately, the reverse is the case in some developing and underdeveloped countries. Engineers in the underdeveloped countries, for instance, rarely practice technological development, possibly due to the high level of poverty in such environments. Another reason that could be advanced for this is the shortage of skilled manpower in the engineering profession that could champion technological breakthrough, similar to the channels operated by the world economic powers. The technological development of nations could be enhanced by the formulation of active research teams. Such teams should be focused on solving practical industrial problems. With government funding, developed countries could create an enabling environment and infrastructure to facilitate the participation of engineers, conceptually, in international projects. Unfortunately, many developing countries cannot afford to fund such international interactions for many of their engineers. However, digital tools and techniques of today are making it easier for international participation via remote means, improving economies everywhere.

Challenges on a community may be viewed from the perspective of the problems faced by the inhabitants of that community. Challenges faced by a community could be local or international. Local challenges suggest the need can be satisfied by that community. This need may not be relevant to other communities; for example, an ISE may be in a position to advise the local government chairman of a community on the disbursement of funds, under the control of the local government, on roads. Decision science models, a systems approach tool, could be used to prioritize certain criteria such as the number of users, the economic index of the various towns and villages, the level of business activities, the number of active industries, the length of the road, and the topography or the shape of the road. In the digital era, access to data, experts, and other resources becomes much easier and enhances the effort through the ISEs efficiently utilizing this access, and thus helps the community to thrive.

Soon after graduation, the industrial and systems engineer is expected to tackle a myriad of problems classified as social, political, and economic. This presents a great challenge for these professionals living in the society where these problems exist. Consider the social problems of electricity generation, water provision, flood control, etc. The ISE in the society where these problems exist is expected to work together with other engineers in order that these problems may be solved. They use the systems approach to design, improve on existing designs, and install integrated systems of men, materials, and equipment in such a way as to optimize the use of resources.

In electricity distribution, the ISE should be able to develop scientific tools for the distribution of power generation as well as the proper scheduling of the maintenance tasks to which the facilities must be subjected, as two examples. As these systems get larger, digital tools are emerging to make improvement easier and quicker. One example in this regard is augmented virtual reality. Another example, hybrid energy, such as solar windmills, is improving impact on the environment while producing electricity in ever higher quantities as demand soars. The distribution network should be such that cost is minimized. Loss prevention is a key factor to consider. The purchased materials for maintenance in these systems should be quality controlled with a minimum acceptable standard allowed. In solving water problems, the primary distribution route is a major concern. The ISE may need to develop reliability models that could be applied to predict the life of components used in the system. Using Internet tools, the ISE can now connect all the constituencies within the system real time, making the system more efficient and reliable.

The ISE works in a wide range of industries, which include manufacturing, logistics, service, and defense. In manufacturing, the ISE must ensure that the equipment, manpower, and other resources in the process are integrated in such a manner so as to maintain efficient operation and ensure continuous improvement. The ISE functions in the logistics industry through the management of supply chain systems (e.g., manufacturing facilities, transportation carriers, distribution hubs, retailers), integrating and optimizing fulfillment of customer orders in the most cost-effective way. In the service industry, the ISE provides consultancy in areas related to organizational effectiveness, service quality, information systems, project management, banking, service strategy, etc. In the defense industry, the ISE provides tools to support the management of military assets and military operations in an effective and efficient manner. The ISE works with a variety of job titles. Typical job titles of ISE graduates include industrial engineer, manufacturing engineer, logistics engineer, supply chain engineer, quality engineer, systems engineer, operations analyst, management engineer, and management consultant.

Research in the United States and other countries show that a high proportion of ISE graduates work in consultancy firms or as independent consultants, helping companies to engineer processes and systems to improve productivity, effect efficient operation of complex systems, and manage and optimize these processes and systems. After a university education, the industrial and systems engineer acquires skills from practical exposure in the industry. Depending on the organization that the ISE works for, the experience may differ in depth or coverage. The trend of professional development in ISE is rapidly changing in recent times. This is enhanced by the ever-increasing development in the information and communication technology (ICT) sector of the economy.

Computers and information systems have changed the way ISEs do business. The unique competencies of an ISE are greatly enhanced by the powers of the computer. Today, the fields of application have widened dramatically, ranging from the traditional areas of production engineering, facilities planning, and material handling, to the design and optimization of more broadly defined systems, operations research (OR), and modeling. The ISE is a versatile professional that uses scientific tools in problem solving through a holistic and integrated approach—a systems approach.

ISEs are the bridge between management and engineering where scientific methods are used heavily in making managerial decisions. The ISE field provides the theoretical and intellectual framework for translating designs into economic products and services; rather than with the fundamental mechanics of design. ISE is vital to solving today's critical and complex systems problems in manufacturing, distribution of goods and services, health care, utilities, transportation, entertainment, and the environment. Because ISE is a process-oriented discipline, it is quite amenable to digital tools. The influence of the digital era on ISE can be seen in the following narratives on selected areas of ISE.

SYSTEMS IMPLEMENTATION REQUIREMENTS

The requirements for the implementation of a sustainable system will include the following elements.

IT Infrastructure

Provide overarching guidance to influence corporate IT improvement investments to enable a robust, secure infrastructure for the enterprise-wide digital campaign.

Integrated Environment for Models and Tools

Provide an integrated digital environment (IDE) of models and tools for collaboration, analysis, and visualization across the functional domains.

Operational Standards, Data, and Systems Architectures

Provide overarching guidance on the use of standardized systems architecture and related standards and datasets for use in an IDE for application at the enterprise and system levels.

Life Cycle Strategies and Processes

Develop life cycle strategies and processes for technology transition, system acquisition, and product support based on life cycle activities from concept development to disposal.

Policy and Guidance

Assess and define the required policy and guidance updates and changes to enable full implementation of the digital transformation.

Workforce and Culture

Workforce culture can drive leveraging the digital era. If the workforce can be brought on board, digital implementations can take root. Drive culture change across the organizational enterprise through training, education, mentoring, and change management strategies, enabling a workforce well versed in a digital engineering framework.

COMPUTER TOOLS IN SYSTEMS IMPLEMENTATION

The foundation of all digital tools and what made the digital era possible is the computer. The impact of computers on engineering is multifaceted. Data analysis by ISEs relies on computers as an important and powerful tool for collecting, recording, retrieving, analyzing simple and complex problems, as well as distributing tremendous masses of information in ISE. It saves countless years of tedious work by the ISEs. The computer removes the necessity for ISEs to monitor and control tedious and repetitive processes. There are several powerful computer programs that can reduce the complexity of solving engineering problems. Despite the importance of computers, its potential is so little explored that its full impact is yet to be realized. The computer has allowed several digital tools to develop.

In the early days of desktop computers (then known as microcomputers), the discipline of IE was the first to embrace using computers for a variety of engineering and administrative functions. The first conference on Computers and Industrial Engineering was organized in Orlando, Florida in 1982 as an initiative of Dr. Gary E. Whitehouse, who was then the Department Chair for Industrial Engineering Department at the University of Central Florida. The first author of this book was a PhD student under Dr. Whitehouse at that time and did contribute as a conference planning assistant for the conference. The conference was so successful that Dr. Whitehouse started writing a monthly column in the Industrial Engineer magazine (now ISE magazine) on the topic of Industrial Engineering Applications of Microcomputers. The conference proceedings of the Computers and Industrial Engineering conference later morphed into the archival journal named *Computers and Industrial Engineering Journal*, which is still in operation today. Thus, IE has been at the forefront of embracing the digital era. In those early days, the first author of this book worked on research projects, conference presentations, and journal publications that promoted computer tools in ISE implementations. Some of those projects are as follows:

"Setting Manufacturing Tolerances by Computer Simulation," 1982
"Introduction to STARC—A Computer-based Project Scheduling Package," 1984
"The Impact of the Computer on Resource Allocation Algorithms," 1985
"Cyclic Computer Modeling of Time Series Realizations," 1986
"Computer-based Graphical Analysis of Mortgage Payments," 1986

The second author of this book, early in her career, participated in a research project by an employer to explore the use of microcomputers in analyzing and summarizing time study analysis in delivering pantyhose to stores. They developed a program in C program language on a microcomputer that was so successful that the program was implemented throughout the company to summarize all-time studies, and the resulting database was linked into the mainframe to use in the enterprise software to schedule deliveries, production, and maintenance.

MATHEMATICAL MODELING

One of the first tools to emerge was mathematical modeling. It is the basis for pursuing optimization in the area of OR, which is a notable subarea of ISE. A model is a simplified representation of a real system or phenomenon. Models are abstractions revealing only the features that are relevant to the real system behavior under study. In ISE, virtually all areas have operational challenges that can be modeled in one form or other. In particular mathematical models are elements, concepts, and attributes of a real system represented by using mathematical symbols. Models are powerful tools for predicting the behavior of the real system by changing some items in the models to know the impact of those changes on the behavior of the real system. They provide frames of reference by which the performance of the real system can be measured. They articulate abstractions thereby enabling us to distinguish between relevant or irrelevant features of the real system. Models are prone to manipulation more easily in a way that the real systems are often not. OR has seen improvement in performance and applicability with the emergence of computational tools and new programming languages. For example, MATLAB® software is very useful because it has many toolkits that are suited for different purposes. Unfortunately, MATLAB and the associated toolkits can be quite expensive for licensees. There is a great open-source alternative called "OCTAVE" that is available for free (https://www.gnu.org/software/octave/index). It is a very capable software tool and it offers a great free alternative to MATLAB. The only downside is that it does not have the toolkits, which is why many OR users still purchase MATLAB software.

STATISTICAL MODELING

Statistics is another foundational aspect to many digital tools. There has been an increase in use of statistics in the ISE field in recent years. For this reason, ISEs are exposed to statistical reasoning in school and early in their careers. ISEs need to understand the basic statistical tools to function in a world that is becoming increasingly dependent on quantitative information. This shows that the interpretation of practical and research results in ISE depends to a large extent on statistical methods. Statistics is used widely in just about every area relevant to these fields. For example, it is utilized as a tool for evaluating economic data in "financial engineering." ISEs also employ statistical techniques to establish quality control systems. This involves detecting an abnormal increase in defects, which reflects equipment causing malfunctioning. The questions of what, how, and when do we apply statistical techniques in practical situations and how to interpret the results are answered in courses on statistics.

OPERATIONS RESEARCH

OR specifically provides the mathematical tools required by ISEs in order to carry out their task efficiently. Its aim is to optimize system performance and predict system behavior using rational decision making to analyze and evaluate complex conditions and systems. It is one of the early digital mathematical techniques that quickly exploited the availability of new tools in the digital era. This area of ISE deals with the application of scientific methods to decision making, especially in the allocation of scarce human resources, money, materials, equipment, or facilities. It covers such areas as mathematical and computer modeling and information technology. It could be applied to managerial decision making in the areas of staff and machine scheduling, vehicle routing, warehouse location, product distribution, quality control, traffic light phasing, and police patrolling. Preventive maintenance scheduling, economic forecasting, design of experiments, power plant fuel allocation, stock portfolio optimization, cost-effective environmental protection, inventory control, and university course scheduling are also some of the other problems that could be addressed by employing OR.

Tools such as mathematics and computer modeling can forecast the implications of various choices and identify the best alternatives. OR methodology is applied to a broad range of problems in both the public and private sectors. These problems often involve designing systems to operate in the most effective way. OR is interdisciplinary and draws heavily on mathematical programs. It exposes ISE graduates to a wide variety of opportunities in areas such as pharmaceuticals, ICT, consulting, financial services, manufacturing, research, logistics and supply chain management, health, among others. These graduates are employed as technical analysts with the potential to move into managerial functions. OR education adopts courses from computer science, engineering management, and other engineering programs, to train students to become highly skilled in the quantitative and qualitative modeling and analysis of a wide range of systems-level decision problems. It relates to productivity, efficiency, and quality. It also impacts the creative utilization of analytical and computational skills in problems—solving problems while increasing the knowledge necessary to become truly competent in today's highly competitive business environment.

OR has tremendous impact on almost every facet of modern life including marketing, oil and gas industry, the judiciary, defense, computer operations, inventory planning, the airline system and international banking, among others. It is a subject of beauty whose application seems endless. Mathematically based OR modeling of problems has been simplified and made more widespread and accessible via digital tools, such as meta-computing, artificial neural network, and machine learning. Another emerging digital tool that OR can capitalize on is the Internet bot, web robot, robot, or simply bot, which is a software application that runs automated tasks over the Internet. Typically, bots perform tasks that are simple and repetitive, much faster than a person could. The most extensive use of bots is for web crawling, in which an automated script fetches, analyzes, and files information from web servers. More than half of all web traffic is generated by bots. With this digital capability, remote computing can be done for systems optimization applications.

INTELLIGENT SYSTEMS

Artificial intelligence (AI) is the most readily acknowledged branch of the fast-emerging digital area of intelligent systems. Although AI is not a direct subarea of IE, its decision-making basis, particularly in the expert systems (ES) area, makes it directly related to the application of IE. Some early AI/ES worked on by this book's first author include the following IE applications:

- Fuzzy Engineering Expert Systems with Neural Network, 2002
- Neural Network as a Simulation Metamodel in Economic Analysis of Risky Projects, 1993
- Taxonomical Analysis of Project Activity Networks Using Competitive Artificial Neural Networks, 2000
- Prediction versus Classification in a Manufacturing Application, 1993
- Taxonomical Analysis of Project Activity Networks Using Competitive Artificial Neural Networks, 2000

The aim of studying AI is to understand how the human mind works, leading to an appreciation of the nature of intelligence, and to engineered systems that exhibit intelligence. Some of the basic keys to understanding intelligence are vision, robotics, and language. Some related aspects to AI include reasoning, knowledge representation, natural language understanding, genetic algorithm, and ES. Studies on reasoning have evolved from the following dimensions: Case-based, non-monotonic, model, qualitative, automated, spatial, temporal, and common sense. For knowledge representation, knowledge bases are used to model application domains and to facilitate access to stored information. Knowledge representations that originally concentrated around protocols were typically tuned to deal with relatively small knowledge bases, but provided powerful reasoning services, and were highly expressive.

Natural language generation systems are computer software systems that produce texts in English and other human languages, often from nonlinguistic input data. Natural language systems, like most AI systems, need substantial amounts of knowledge that is difficult to acquire. In general terms, these problems are due to the complexity, novelty, and poorly understood nature of the tasks involved, and are worsened by the fact that people write in vastly different ways.

A genetic algorithm is a search algorithm based on the mechanics of natural selection and natural genetics. It is an iterative procedure maintaining a population of structures that are candidate solutions to specific domain challenges. During each generation, the structures in the current population are rated for their effectiveness as solutions, and, on the basis of these evaluations, a new population of candidate structure is formed using specific genetic operators such as reproduction, crossover, and mutation.

An ES is computer software that can solve a narrowly defined set of problems using information and reasoning techniques normally associated with a human expert. It could also be viewed as a computer system that performs at or near the level of a human expert in a particular field of endeavor.

These techniques are evolving fast to solve bigger and more complex problems. How to use these techniques will be developed by ISEs and applied to systems problems to enhance the outcomes even more.

HUMAN FACTORS AND ERGONOMICS

Human factors (HF) engineering is a practical discipline dealing with the design and improvement of productivity and safety in the workplace. It is concerned with the relationship of manufacturing and service technologies interacting with humans. The focus is not restricted to manufacturing only but also to service systems. The main methodology of ergonomics involves the adaptation of the components of human-machine-environment systems by means of human-centered design of machines and the work environment in production systems. Ergonomics studies human perceptions, motions, workstations, machines, products, and work environments.

Today's ever-increasing concerns about humans in the technological domain make this field very appropriate. People in their everyday lives or in carrying out their work activities create products and environments for use. In many instances, the nature of these products and environments directly influences the extent to which they serve their intended human use. The discipline of HF deals with the problems and processes that are involved in human's efforts to design these products and operating environments, such that the products optimally serve their intended use by humans without harm to the users. This general area of human endeavor (and its various facets) has come to be known as HF engineering, or simply HF, biomechanics, engineering psychology, or ergonomics. HF and ergonomics have been traditionally manual based in physical experiments. Today, digital tools are helping to enhance the work of ISEs in this area. For example, high-speed video and video analysis have enhanced the analysis of jobs. Creating standardized training videos for operating procedures has allowed operators to review and follow guidelines while on the production line.

The purpose of ISE is, in general, to seek an optimal solution under given conditions. As one of the main areas of ISE, HF engineering pursues the same goal. The only difference is the target, which is people. HF or ergonomics have become familiar terms, as can be seen from the use of ergonomics in applications from simple tools to very sophisticated airplanes. The study of HF has as its goal maximizing human capacity, usability, and comfort while minimizing human errors, accidents, and injury. Therefore, in ergonomics, we design a system that takes into account human capabilities and skills while optimizing technology and human interactions.

The design of products without the use of HF input can cause loss of productivity and sometimes the loss of lives. Everyday life is affected by this unsatisfactory approach, as is evidenced by the complexity of video cassette recorder (VCR) programming tasks on remote controllers, automobile diagnostic repair problems, or even the setup and use of business or personal computers. A well-known example frequently mentioned with regard to a lack of HF input was the nuclear power plant accident at Three Mile Island on March 28, 1979. After the accident, the US Nuclear Regulatory Commission (NRC) published a summary report. The NRC stated in the report:

> The accident at the Three Mile Island Unit 2 (TMI-2) nuclear power plant near Middletown, Pennsylvania, on March 28, 1979, was the most serious in U.S. commercial nuclear power plant operating history, even though it led to no deaths or injuries to plant workers or members of the nearby community. But it brought about sweeping changes involving emergency response planning, reactor operator training, **human factors engineering**, radiation protection, and many other areas of nuclear power plant operations. ...... The accident began about 4:00 a.m. on March 28, 1979, when the plant experienced a failure in the secondary, non-nuclear section of the plant. The main feedwater pumps stopped running, caused by either a mechanical or electrical failure, which prevented the steam generators from removing heat. First the turbine, then the reactor automatically shut down. Immediately, the pressure in the primary system (the nuclear portion of the plant) began to increase. In order to prevent that pressure from becoming excessive, the pilot-operated relief valve (a valve located at the top of the pressurizer) opened. The valve should have closed when the pressure decreased by a certain amount, but it did not. *Signals available to the operator failed to show that the valve was still open.* As a result, cooling water poured out of the stuck-open valve and caused the core of the reactor to overheat. As coolant flowed from the core through the pressurizer, the instruments available to reactor operators provided confusing information. There was no instrument that showed the level of coolant in the core. Instead, the operators judged the level of water in the core by the level in the pressurizer, and since it was high, *they assumed that the core was properly covered with coolant*. In addition, there was no clear signal that the pilot-operated relief valve was open. As a result, as alarms rang and warning

> lights flashed, the operators did not realize that the plant was experiencing a loss-of-coolant accident. They took a series of actions that made conditions worse by simply reducing the flow of coolant through the core.
>
> (http://www.nrc.gov/reading-rm/doc-collections/fact-sheets/3mile-isle.html *for details)*

After the accident at Three Mile Island, an awareness of the importance of good design has been acknowledged in many areas including product design. Multiple terms are used to describe the skills applied to the designing and developing of systems and their products so that the results are user-centered, not equipment-centered. In addition to these design issues, the study of HF has made many contributions to safety issues. In the early era of HF research, productivity improvement was the main topic but as technology has advanced, especially computer technology, the usability of tools, machine, and computer software has been placed in the main core of research topics. In the modern digital era, digital and automated tools are available to aid the work of HF and ergonomics.

Design for adjustability is typically used for equipment or facilities that can be adjusted to fit a wide range of individuals. Chairs, vehicle seats, steering columns, and tool supports are devices that are typically adjusted to accommodate the worker population ranging from 5th percentile females to 95th percentile males. Obviously, designing for adjustability is the preferred method of design, but there is a trade-off with the cost of implementation. Fortunately, new digital design techniques are helping to mitigate the limitations of traditional approaches to HF and ergonomics. One exciting emerging digital tool in this respect is the development of digital humanoids or digital humans, which are designed with human-like attributes for HF studies and experiments.

MANUFACTURING SYSTEMS

In order to survive in the competitive environment we are in today, significant changes need to be made in the ways manufacturing organizations design, make, sell, and service their products. Manufacturers are committed to continuous improvement in product design, defect levels, and costs. Today this is achieved by fusing the design, manufacturing, distribution, and marketing into a complete whole—using a systems approach.

Manufacturing systems consist of two parts: Science and automation. Manufacturing science refers to investigations into the processes involved in the transformation of raw materials into finished products. Traditionally, manufacturing science refers to the techniques of work-study, inventory systems, material requirement planning, Toyota system, etc. On the other hand, the automation aspect of manufacturing covers issues like e-manufacturing, the use of computer-assisted manufacturing systems (numerical control [NC], computer numerical control [CNC], and direct numerical control [DNC]), automated material handling systems, group technology, automated manufacturing processes and quality monitoring, flexible manufacturing systems, process planning and control, etc.

ISE students research into manufacturing in combination with courses in finance, manufacturing processes, and personnel management. They also engage in manufacturing design projects. This exposes the students to a manufacturing environment with activities in the design or improvement of manufacturing systems, product design, and quality.

HUMAN-COMPUTER INTERACTION

The website InternetWorldStats.com estimated in 2020 that 2,405,518,376 people, or 34.3% of the world's population, are Internet users, meaning they have access to the Internet either through a computer or mobile phone. Computer systems, including both hardware and software, have become more and more sophisticated. End users have different levels of skills and knowledge. But regardless of their skill and knowledge levels, end users have a single goal in common: To complete the given tasks as soon as possible or find what they want quickly when surfing the Internet. However, most do not think about how their use of these tools may be injuring them, as the injury happens over time. Better design for use of these tools must be found. But the design has to work for everyone, which can be challenging. For example, consider a voice recognition system. Many companies are now providing voice response systems rather than a touch-tone response system. Some customers will find it convenient while some will find it difficult. The voice recognition technology must be advanced enough for users not to worry about the question, "How can my speech be perfectly recognized through the phone?" But, for those who have a strong accent or who are beginning to learn English, the voice recognition system may not be favorable.

Human-computer interaction (HCI) is a relatively young field that is still developing compared to other research fields. There is a wide range of people interested in HCI who are in linguistics, social and behavioral science, computer engineering, information technology, and so on.

There are three major topics in HCI:

1. Understanding user characteristics.
2. Providing design principles and guidelines.
3. Conducting usability tests.

HCI professionals, including HF specialists, collect user data and give user-characteristics information to the system developers so that the developers know more about what the user wants and how the user behaves. The HCI professionals also provide the user more user-friendly support materials with help from system developers.

User-friendly is a very familiar word in HCI. It indicates that consumers now prefer software packages or devices having user-friendly features. However, considering the nature of software that ranges from performing very simple functions such as basic calculations to extremely complex functions such as control of a nuclear power plant, it is not easy to define *user-friendliness*. Also, as products become more complex, the necessary interfaces will likely have numerous displays, menus, display formats, control systems, and many levels of interface functions. The trend toward a greater number of functions is an important problem because the additional functions make the interface more complex and increase the number of choices the user must make, thereby decreasing user-friendliness. Any product that is designed without paying attention to the user often leaves a huge gap between the user's capabilities and the expectations the product places on the user. The goal of the HF specialist is to help narrow this gap by understanding and paying attention to the needs of the user rather than what the system developers want to make, and helping the system developers understand the user needs.

DEJI SYSTEMS MODEL FOR OPERATIONAL EXCELLENCE

Based on the foregoing discussions, it is essential to have a systems tool that can be implemented in a digital environment to actualize the input-process-output framework presented at the beginning of this chapter. The DEJI Systems Model®, a trademarked systems tool for design, evaluation, justification, and integration of any operational system (Badiru, 2014, 2019) is designed to do this. Figure 3.2 illustrates the schematic of the DEJI Systems Model. Most of the typical operational factors in operational excellence are covered in the model. Of greatest importance in the DEJI Systems Model is the integration

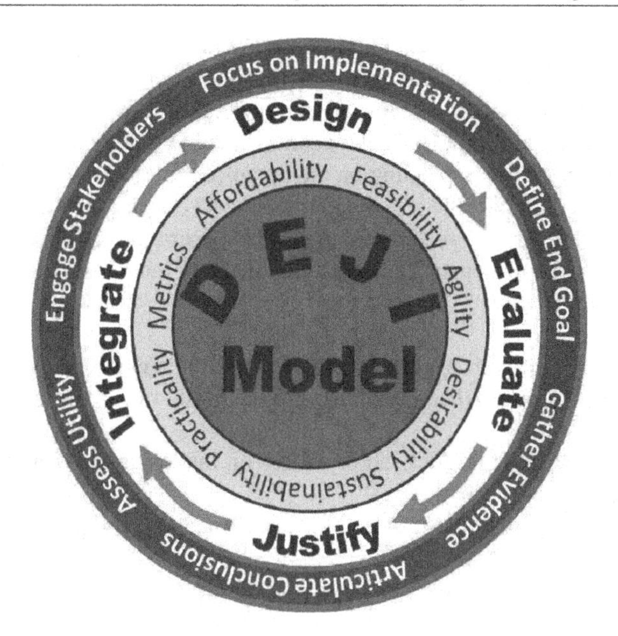

FIGURE 3.2 DEJI Systems Model application to operational excellence.

stage. Integration ensures that all elements that are essential for operational excellence are included in the implementation phase of any process.

Systems quality is at the intersection of efficiency, effectiveness, and productivity. Efficiency provides the framework for quality in terms of resources and inputs required to achieve the desired level of quality. Effectiveness comes into play with respect to the application of product quality to meet specific needs and requirements of an organization. Productivity is an essential factor in the pursuit of quality as it relates to the throughput of a production system. To achieve the desired levels of quality, efficiency, effectiveness, and productivity, a new research framework must be adopted. The DEJI model is relevant for research efforts in quality engineering and technology applications and other SE applications.

The DEJI model of SE provides one additional option for SE development applications. Although the model is generally applicable in all types of systems modeling, systems quality is specifically used to describe how the DEJI model is applied. The core stages of the DEJI model are as follows:

- Design
- Evaluation
- Justification
- Integration

Design encompasses any system initiative providing a starting point for a project. Thus, design can include technical product design, process initiation, and concept development. In essence, we can say that "design" represents requirements and specifications. Evaluation can use a variety of metrics both qualitative and quantitative, depending on the organization's needs. Justification can be done based on monetary, technical, or social reasons. Integration needs to be done with respect to the normal or standard operations of the organization. All the operational elements embedded in the DEJI model can be summarized as presented below:

- Design embodies agility, define end goal, and engage stakeholder
- Evaluate embodies feasibility, metrics, gather evidence, and assess utility
- Justify embodies desirability, focus on implementation, and articulate conclusions
- Integrate embodies affordability, sustainability, and practicality

For application purposes, these elements interface and interact systematically to enhance overall operational performance of organizations, particularly in the digital era. In summary, digitization is changing many human processes. For example, autonomous robots (e.g., digital humans) are imparting new possibilities in the banking, health care, industry, and service sectors. This development will continue to create new avenues for operational excellence. For example, the company name Boston Dynamics has been engrossed in developing special dynamic robots that exhibit human-like physical characteristics and capabilities. One particular robot is designed to expedite package loading and unloading in an industrial supply chain.

REFERENCES

Badiru, Adedeji B. (2014), "Quality Insights: The DEJI® Model for Quality Design, Evaluation, Justification, and Integration," *International Journal of Quality Engineering and Technology*, Vol. 4, No. 4, pp. 369–378.

Badiru, Adedeji B. (2019), *Systems Engineering Models: Theory, Methods, and Applications*, Taylor & Francis Group/CRC Press, Boca Raton, FL.

Oke, S. A. (2014), "An Overview of Industrial and Systems Engineering," in Badiru, A. B., editor, *Handbook of Industrial and Systems Engineering*, 2nd edition, CRC Press, Boca Raton, FL, Chapter 10, pp. 185–196.

Work Performance Excellence via Project Management

4

INTRODUCTION

The emergence of the digital era has created new opportunities to advance work, particularly where people are involved. Using a project management framework, this chapter presents tools and techniques for teamwork, work simplification, group decision making, and other related topics. Multinational projects, particularly, pose unique challenges pertaining to reliable power supply, efficient communication systems, credible government support, dependable procurement processes, consistent availability of technology, progressive industrial climate, trustworthy risk mitigation infrastructure, regular supply of skilled labor, uniform focus on quality of work, global consciousness, hassle-free bureaucratic processes, coherent safety and security systems, steady law and order, unflinching focus on customer satisfaction, and fair labor relations. Assessing and resolving concerns about these issues in a step-by-step fashion using project management will create a foundation of success for a large project. While no system can be perfect and satisfactory in all aspects, a tolerable trade-off of the factors is essential for project success.

Work processes are the basis for accomplishing organizational goals and objectives. If work can be managed better, leveraging tools of the digital era,

DOI: 10.1201/9781003052036-4

operational excellence can be assured. In the context of organizational needs, work boils down to the pursuit of one or more of the following:

1. Achievement of a required end product.
2. Provision of a needed service.
3. Realization of a desired result.

A work package is a combination of activities and tasks. The work packages constituting the above goals and objectives have been augmented by the digital era in recent years. However, we need to articulate the tools, techniques, models, and framework for leveraging the digital era in pursuit of operational excellence. As General Hap Arnold said during WWII, we should leverage and use new tools and techniques that appear within our operating realm:

> This method of using officers & civilians for purely analytical work has proven fruitful in many fields, & the Army Air Forces should make use of it.
>
> *General Hap Arnold, Oct 1942*

This quote is still relevant today and points to the need to leverage the new tools of the digital era to facilitate work. From a project management viewpoint, work packages should be managed like any project, with a focus on what, who, when, where, why, and how. The "how" aspect of managing work packages is where the application of digital tools is most noticeable. Project management is work management and vice versa. If not managed like a project, some work elements run the risk of being a means to no end. Enterprise software, one of the early tools of the digital era, has taken this idea of running work packages like a project, with all the basic functions, listed below, of a project built into the programming. These software also use the hierarchical levels of project management to create reports and do analysis at varying levels of detail.

I. A. Badiru (2016) comments that "It is our natural biological imperative to work. Unfortunately, there is no imperative for that work to be useful." Project management is the process of managing, allocating, and timing resources in order to achieve a given objective in an expedient manner. It is the process of achieving objectives by utilizing the combined capabilities of available resources. The objective may be stated in terms of time (schedule), performance output (quality), or cost (budget). Time is often the most critical aspect of managing any project. Time is the physical platform over which project accomplishments are made. Therefore, it must be managed concurrently with all other important aspects of any project. Project management covers the following basic functions:

1. Planning
2. Organizing

3. Scheduling
4. Control

The complexity of a project can range from simple, such as the painting of a vacant room, to very complex, such as the introduction of a new high-tech product. The technical differences between project types are of great importance when selecting and applying project management techniques.

Project management techniques are widely used in many human endeavors, such as construction, banking, manufacturing, marketing, health care, sales, transportation, research and development, academic, legal, political, and government establishments, just to name a few. In many situations, the on-time completion of a project is of paramount importance. Delayed or unsuccessful projects not only translate to monetary losses but also impede subsequent undertakings. Project management takes a hierarchical view of a project environment, covering the top-down levels shown below:

1. System level
2. Program level
3. Project level
4. Task level
5. Activity level

PROJECT REVIEW AND SELECTION FOR WORK PERFORMANCE

Project selection is an essential first step in focusing the efforts of an organization. Figure 4.1 presents a simple graphical evaluation of project selection, whereby work simplification is a factor of interest. The vertical axis represents the value-added basis of the project under consideration while the horizontal axis represents the level of complexity associated with the project. In this example, value can range from low to high while complexity can range from easy to difficult. The figure shows four quadrants containing regions of high value with high complexity, low value with high complexity, high value with low complexity, and low value with low complexity. A fuzzy region is identified with an overlaid circle. The organization must evaluate each project on the basis of the organization's value streams. If a team is formed around a task, an important question is whether or not value is added by the team. The figure can be modified to represent other factors of interest to an organization instead of value-added and project complexity.

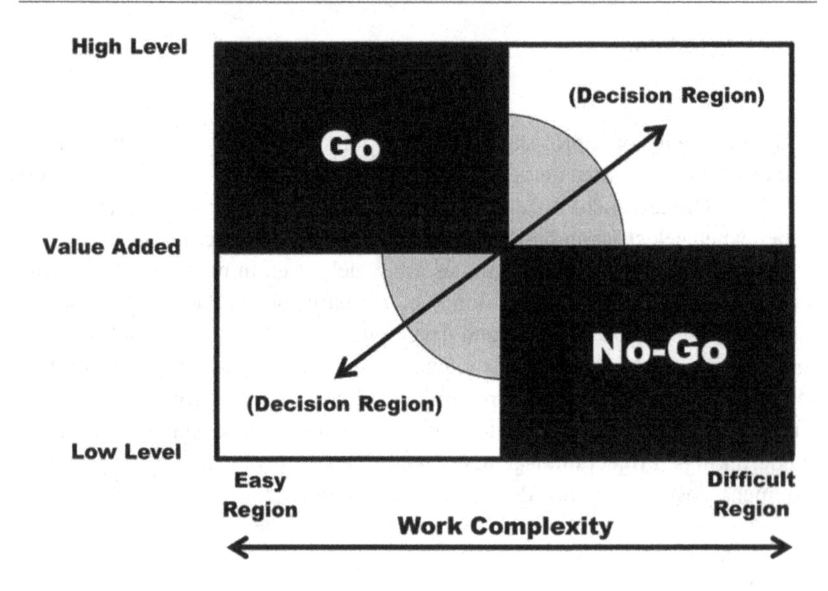

FIGURE 4.1 Project selection with respect to work simplification.

In some cases, work packages may consist of different sets of activities or tasks. A task that is in one work package during a particular work cycle may not be in the work package in another cycle. As an example, a task may be assigned as "going to the bank to make a deposit." Individual activities within that task may include signing the check, determining the bank address, and getting in the car for the drive to the bank. For the purpose of work optimization or work simplification, some of the elements of a work package may be altered, removed, reassigned, redesigned, or modified. It all depends on the prevailing operational scenarios and current needs of the organization. In each case, digital tools may be available for the analysis and decision making related to work packages. A good example is the use of a digital GPS (global positioning system) to determine the location of the bank rather than using a traditional printed map.

SIZING OF PROJECTS

Associating a size measure to an industrial project provides a means of determining level of effort required. A simple guideline is presented below, but should be modified to fit the organization's own environment:

- Major (over 60 man-months of effort)
- Intermediate (6–60 man-months)
- Minor (less than 6 man-months)

By understanding the size of the project in terms of effort needed, the decision to proceed will take into consideration the resources needed, and the resources can be properly allocated.

Planning Levels

When selecting projects and associated work packages, planning should be done in an integrative and hierarchical manner following the levels of planning presented below:

- Supra level
- Macro level
- Micro level

Understanding at the supra level provides the context in which the project will run. Which key performance indicators (KPI) are in most need of improvement? Does the selected work impact the company's goals or a KPI? Then coming down to the macro level allows insight into the overall resources needed and time frame required, using the sizing information. It can be an iterative effort between supra- and macro levels until a project is selected. Once a selected and scoped out, the micro-level planning can be done.

Project Alert Scale: Red, Yellow, Green Convention

Hammersmith (2006) presented a guideline for an alert scale for project tracking and evaluation. He suggested putting projects into categories of RED, YELLOW, or GREEN with the definitions below:

RED (if not corrected, project will be late and/or be over budget)
YELLOW (project is at risk of turning RED)
GREEN (project is on time and on budget)

Defining the trigger for each level and communicating this to the team and to oversight management greatly improves project performance. Everyone is aware of what good performance is and when to act to correct the direction and performance of the project.

Body of Knowledge Methodology

The general body of knowledge (PMBOK®) for project management is published and disseminated by the Project Management Institute (PMI). It is an excellent source of project management techniques. The body of knowledge comprises specific knowledge areas, which are organized into the following broad areas:

1. Project **integration** management
2. Project **scope** management
3. Project **time** management
4. Project **cost** management
5. Project **quality** management
6. Project **human** resource management
7. Project **communications** management
8. Project **risk** management
9. Project **procurement** management

The above segments of the body of knowledge of project management cover the range of functions associated with a typical project, particularly complex ones. The body of knowledge continues to expand to be more responsive to new performance processes facilitated by the digital era. There are many software packages available for managing projects, enhanced by excellent tracking communication tools afforded with today's technology.

Project Systems Structure

The overall project management system structure can be outlined as summarized below.

Problem identification

Problem identification is the stage where a need for a proposed project is identified, defined, and justified. A project may be concerned with the development of new products, implementation of new processes, or improvement of existing facilities.

Project definition

Project definition is the phase at which the purpose of the project is clarified. A mission statement is the major output of this stage. For example, a prevailing low level of productivity may indicate a need for a new manufacturing technology. In general, the definition should specify how project management will be used to avoid missed deadlines, poor scheduling, inadequate resource allocation, lack of coordination, poor quality, and conflicting priorities.

Project planning

A plan represents the outline of the series of actions needed to accomplish a goal. Project planning determines how to initiate a project and execute its objectives. It may be a simple statement of a project goal or it may be a detailed account of procedures to be followed during the project. Planning can be summarized as follows:

- Objectives
- Project definition
- Team organization
- Performance criteria (time, cost, quality)
- Schedule

Project organizing

Project organization specifies how to integrate the functions of the personnel involved in a project. Organizing is usually done concurrently with project planning. Directing is an important aspect of project organization. Directing involves guiding and supervising the project personnel. It is a crucial aspect of the management function. Directing requires skillful managers who can interact with subordinates effectively through good communication and motivation techniques. A good project manager will facilitate project success by directing his or her team through proper task assignments and coaching, toward the project goal.

Team members perform better when there are clearly defined expectations. They need to know how their job functions contribute to the overall goals of the project. Team members should be given some flexibility for self-direction in performing their functions. Individual team member needs and limitations should be recognized by the manager when directing project functions. Organizing and directing a project requires skills dealing with motivating, supervising, and delegating.

Resource allocation

Project goals and objectives are accomplished by allocating resources to functional requirements. Resources can consist of money, people, equipment, tools, facilities, information, skills, and so on. These are usually in short supply. The people needed for a particular task may be committed to other ongoing projects. A crucial piece of equipment may be under the control of another team. Using software to manage resources has greatly improved the coordination and communication among team members and their managers.

Project scheduling

Timeliness is the essence of project management. Scheduling is often the major focus in project management. The main purpose of scheduling is to define the activities needed and allocate resources so that the overall project objectives are achieved within a reasonable time span. Project objectives are often conflicting in nature. For example, minimization of the project completion time and minimization of the project cost can be conflicting objectives. That is, one objective may be improved at the expense of the other objective. Therefore, project scheduling is a multiple-objective decision-making problem, and the tools available today, even beyond the project management software, are many.

In general, scheduling involves the assignment of time periods to specific tasks within the work schedule. Resource availability, time limitations, urgency level, required performance level, precedence requirements, work priorities, technical constraints, and other factors complicate the scheduling process. Thus, the assignment of a time slot to a task does not necessarily ensure that the task will be performed satisfactorily in accordance with the schedule. Consequently, careful oversight and control must be developed and maintained throughout the project scheduling process. This used to be done with pencil and paper. Computers and spreadsheets enhanced the ability to schedule more complex projects. Today, software can take the inputs, resources, and constraints, and produce a Program Evaluation Review Technique (PERT) or Gantt chart and calculate the bottlenecks faster than any manual process. Communication of the schedule and performance to that schedule is essential to success.

Project tracking and reporting

This phase involves checking whether project results conform to project plans and performance specifications. Tracking and reporting are prerequisites for project control. A properly organized report of the project status will help identify any deficiencies in the progress of the project and help pinpoint corrective actions. In a manual system, the time to gather, analyze, and report performance could lag behind the project and corrective actions could be taken too late, jeopardizing the success of the project. Today, the automated data collection through reporting inputs from team members and the communication of issues that the software packages provide has reduced this chance for corrective actions being too late.

Project control

Project control requires that appropriate actions be taken to correct unacceptable deviations from expected performance. Control is actuated through measurement, evaluation, and corrective action. Measurement is the process of measuring the relationship between planned performance and actual performance with respect to project objectives. The variables to be measured, the measurement scales, and the measuring approaches should be clearly specified during the planning stage. A communication responsibility matrix and a decision matrix are needed to ensure everyone understands who can determine the action needed, and avoid confusion and duplication of effort.

Project management software is only now being enhanced to take in automated data. Coordinated and diligent reporting of data, or use of electronic means to collect and measure, is needed to have good information to make corrective action decisions. Corrective actions may involve rescheduling, reallocation of resources, or expedition of task performance. Control involves the following:

- Tracking and reporting
- Measurement and evaluation
- Corrective action (plan revision, rescheduling, updating)

Project termination

Termination is the last stage of a project. The phase out of a project is as important as its initiation. The termination of a project should be implemented expeditiously. A project should not be allowed to drag on after the expected completion time. A terminal activity should be defined for a project during the planning phase without review and approval of an agreed upon extended completion date, and communication to all stakeholders. An example of a terminal activity may be the submission of a final report, the power-on of new equipment, or the signing of a release order. The conclusion of such an activity should be viewed as the completion of the project. Arrangements may be made for follow-up activities that may improve or extend the outcome of the project. These follow-up or spinoff projects should be managed as new projects but with proper input-output relationships within the sequence of projects.

WORK PERFORMANCE IN SYSTEMS DECISION ANALYSIS

System decision analysis facilitates a proper consideration of the essential elements of decisions in a project system environment. Essential elements include the problem statement, information, performance measurement, decision model, and an implementation plan of the decision. The recommended steps are as follows.

Step 1: Problem Statement

Problem definition is very crucial. In many cases, *symptoms* of a problem are more readily recognized than its *cause* and *location*. Even after the problem is accurately identified and defined, a benefit/cost analysis may be needed to determine if the cost of solving the problem is justified. A problem involves choosing between competing, and probably conflicting, alternatives.

The components of problem-solving in project management include the following:

- Describing the problem (goals, performance measures)
- Defining a model to represent the problem
- Solving the model
- Testing the solution
- Implementing and maintaining the solution

Step 2: Data and Information Requirements

Information is the driving force for the project decision process. Information clarifies the relative states of past, present, and future events. The collection, storage, retrieval, organization, and processing of raw data are important components for generating information. Without data, there can be no information. Without good information, there cannot be a valid decision. The essential requirements for generating information are as follows:

- Ensuring that an effective data collection procedure is followed
- Determining the type and the appropriate amount of data to collect
- Evaluating the data collected with respect to information potential
- Evaluating the cost of collecting the required data

For example, suppose a manager is presented with a recorded fact that says, "*Sales for the last quarter are 10,000 units.*" This constitutes ordinary data.

There are many ways of using the above data to make a decision depending on the manager's value system. An analyst, however, can ensure the proper use of the data by transforming it into information, such as, "*Sales of 10,000 units for last quarter are within x percent of the targeted value.*" This information is more useful to the manager for decision making than the raw data it comes from.

Step 3: Performance Measures

A performance measure for the competing alternatives should be specified. The decision maker assigns a perceived value to the available alternatives. Setting measures of performance is crucial to the process of defining and selecting alternatives. Some performance measures commonly used in project management are project cost, completion time, resource usage and availability, and stability in the workforce.

Step 4: Decision Model

A decision model provides the basis for the analysis and synthesis of information and is the mechanism by which competing alternatives are compared. To be effective, a decision model must be based on a systematic and logical framework for guiding project decisions. A decision model can be a verbal, graphical, or mathematical representation of the ideas in the decision-making process. A project decision model should have the following characteristics:

- Simplified representation of the actual situation
- Explanation and prediction of the actual situation
- Validity and appropriateness
- Applicability to similar problems

The formulation of a decision model involves three essential components:

- Abstraction: Determining the relevant factors
- Construction: Combining the factors into a logical model
- Validation: Assuring that the model adequately represents the problem

The basic types of decision models for project management are as follows:

- Descriptive
- Prescriptive
- Predictive
- Satisficing
- Optimization

In many situations, two or more of the above models may be involved in the solution of a problem. For example, a descriptive model might provide insights into the nature of the problem; an optimization model might provide the optimal set of actions to take in solving the problem; a satisficing model might temper the optimal solution with reality; a prescriptive model might suggest the procedures for implementing the selected solution; and a predictive model will test the outcome of a given solution.

Step 5: Making the Decision

Using the available data, information, and the decision model, the decision maker will determine the real-world actions that are needed to solve the stated problem. A sensitivity analysis may be useful for determining what changes in parameter values might cause a change in the decision.

Step 6: Implementing the Decision

A decision represents the selection of an alternative that satisfies the objective stated in the problem statement. A good decision is useless until it is implemented. An important aspect of a decision is to specify how it is to be implemented. Selling the decision and the project to management requires a well-organized persuasive presentation. The way a decision is presented can directly influence whether or not it is adopted. The presentation of a decision should include at least the following: an executive summary, technical aspects of the decision, managerial aspects of the decision, resources required to implement the decision, cost of the decision, the time frame for implementing the decision, and the risks associated with the decision.

GROUP DECISION MAKING

While there are many new techniques developed for computer decision making, some of the best decisions are still those derived from group discussion and participation. Group decision making is a key part of project planning and control. Systems decisions are often complex, diffuse, and distributed, particularly in a digital environment. No one person has all the information to make all decisions accurately. As a result, crucial decisions are made by a group of people. Some organizations use outside consultants with appropriate expertise to make recommendations for important decisions. Other organizations set up their own internal consulting groups without having to go outside the organization. Decisions can be made through linear responsibility, in which case one person makes the final decision based on inputs from other people. Decisions can also be made through shared responsibility, in which case a group of people share the responsibility for making joint decisions. The major advantages of group decision making are listed below:

1. Facilitation of a systems view of the problem environment.
2. Ability to share experience, knowledge, and resources. Many heads are better than one. A group will possess greater collective ability to solve a given decision problem.
3. Increased credibility. Decisions made by a group of people often carry more weight in an organization.
4. Improved morale. Personnel morale can be positively influenced because many people have the opportunity to participate in the decision-making process.
5. Better rationalization. The opportunity to observe other people's views can lead to an improvement in an individual's reasoning process.
6. Ability to accumulate more knowledge and facts from diverse sources.
7. Access to broader perspectives spanning different problem scenarios.
8. Ability to generate and consider alternatives from different perspectives.
9. Possibility for a broad-based involvement, leading to a higher likelihood of support.
10. Possibility for group leverage for networking, communication, and political clout.

Despite the much-desired advantages, group decision making does have risks. Some possible disadvantages of group decision making are listed below:

1. Difficulty in arriving at a decision.
2. Slow operating time frame.
3. Possibility for individuals conflicting views and objectives to delay the decision.
4. Reluctance of some individuals in to implement the decision if they didn't agree to it.
5. Potential for power struggles and conflicts among the group.
6. Loss of productive employee time.
7. Too much compromise may lead to less than optimal group output.
8. Risk of one individual dominating the group.

9. Over-reliance on group process may impede agility of management to make decisions fast.
10. Risk of dragging feet due to repeated and iterative group meetings.

For major decisions and long-term group activities, arrange for team training that allows the group to learn the decision rules and responsibilities together.

Brainstorming

Brainstorming is a way of generating many new ideas. In brainstorming, the decision group comes together to discuss alternate ways of solving a problem. The members of the brainstorming group may be from different departments, may have different backgrounds and training, and may not even know one another. The diversity of the participants helps create a stimulating environment for generating different ideas from different viewpoints. The technique encourages free outward expression of new ideas no matter how far-fetched the ideas might appear. No criticism of any new idea is permitted during the brainstorming session. A major concern in brainstorming is that extroverts may take control of the discussions. For this reason, an experienced and respected individual should manage the brainstorming discussions. The group leader establishes the procedure for proposing ideas, keeps the discussions in line with the group's mission, discourages disruptive statements, and encourages the participation of all members.

Delphi Method

The traditional approach to group decision making is to obtain the opinion of experienced participants through open discussions. An attempt is made to reach a consensus among the participants. However, open group discussions are often biased because of the influence of subtle intimidation from dominant individuals. Even when the threat of a dominant individual is not present, opinions may still be swayed by group pressure. This is called the "bandwagon effect" of group decision making.

The Delphi method attempts to overcome these difficulties by requiring individuals to present their opinions anonymously through an intermediary. The method differs from the other interactive group methods because it eliminates face-to-face confrontations. It was originally developed for forecasting applications, but it has been modified in various ways for application to different types of decision making. The method can be quite useful for project management decisions. It is particularly effective when decisions must be based on a broad set of factors.

Nominal Group Technique

The nominal group technique is a silent version of brainstorming. It is a method of reaching consensus. Rather than asking people to state their ideas aloud, the team leader asks each member to jot down a minimum number of ideas, for example, five or six. A single list of ideas is then written on a chalkboard for the whole group to see. The group then discusses the ideas and weeds out some iteratively until a final decision is made. The nominal group technique is easier to control. Unlike brainstorming where members may get into shouting matches, the nominal group technique permits members to silently present their views. In addition, it allows introversive members to contribute to the decision without the pressure of having to speak out too often.

In all of the group decision-making techniques, an important aspect that can enhance and expedite the decision-making process is to require that members review all pertinent data before coming to the group meeting. This will ensure that the decision process is not impeded by trivial preliminary discussions. Some disadvantages of group decision making are as follows:

1. Peer pressure in a group situation may influence a member's opinion or discussions.
2. In a large group, some members may not get to participate effectively in the discussions.
3. A member's relative reputation in the group may influence how well his or her opinion is rated.
4. A member with a dominant personality may overwhelm the other members in the discussions.
5. The limited time available to the group may create a time pressure that forces some members to present their opinions without fully evaluating the ramifications of the available data.
6. It is often difficult to get all members of a decision group together at the same time.

Interviews, Surveys, and Questionnaires

Interviews, surveys, and questionnaires are important information gathering techniques. They also foster cooperative working relationships. They encourage direct participation and inputs into project decision-making processes. They provide an opportunity for employees at the lower levels of an organization to contribute ideas and inputs for decision making. The greater the number of people involved in the interviews, surveys, and questionnaires, the more valid the final decision.

Multivote

Multivoting is a series of votes used to arrive at a group decision. It can be used to assign priorities to a list of items. It can be used at team meetings after a brainstorming session has generated a long list of items. Multivoting helps reduce such long lists to a few items, usually three to five.

Despite the noted disadvantages, group decision making definitely has many advantages that may nullify the shortcomings. The advantages as presented earlier will have varying levels of effect from one organization to another. The Triple C approach (Badiru, 2008, Badiru et al., 2008) may also be used to improve the success of decision teams. Teamwork can be enhanced in group decision making by adhering to the following guidelines:

1. Get a willing group of people together.
2. Set an achievable goal for the group.
3. Determine the limitations of the group.
4. Develop a set of guiding rules for the group.
5. Create an atmosphere conducive to group synergism.
6. Identify the questions to be addressed in advance.
7. Plan to address only one topic per meeting.

HIERARCHY OF WORK PACKAGES

The traditional concepts of systems analysis are applicable to project management. The definitions of a project system and its components are as follows.

System

A project system consists of interrelated elements organized for the purpose of achieving a common goal. The elements are organized to work synergistically to generate a unified output that is greater than the sum of the individual outputs of the components.

Program

A program is a very large and prolonged undertaking. Such endeavors often span several years. Programs are usually associated with particular systems. For example, we may have a space exploration program within a national defense system.

Project

A project is a time-phased effort of much smaller scope and duration than a program. Programs are sometimes viewed as consisting of a set of projects. Government projects are often called *programs* because of their broad and comprehensive nature. Industry tends to use the term *project* because of the short-term and focused nature of most industrial efforts.

Task

A task is a functional element of a project. A project is composed of a sequence of tasks that all contribute to the overall project goal.

Activity

An activity can be defined as a single element of a project. Activities are generally smaller in scope than tasks. In a detailed analysis of a project, an activity may be viewed as the smallest, practically indivisible work element of the project. For example, we can regard a manufacturing plant as a system. A plantwide endeavor to improve productivity can be viewed as a program. The installation of a flexible manufacturing system is a project within the productivity improvement program. The process of identifying and selecting equipment vendors is a task, and the actual process of placing an order with a preferred vendor is an activity.

The emergence of systems development has had an extensive effect on project management in recent years. A system can be defined as a collection of interrelated elements brought together to achieve a specified objective. In a management context, the purpose of a system is to develop and manage operational procedures and to facilitate an effective decision-making process. Some of the common characteristics of a system include the following:

1. Interaction with the environment
2. Objective
3. Self-regulation
4. Self-adjustment

Representative components of a project system are the organizational subsystem, planning subsystem, scheduling subsystem, information management subsystem, control subsystem, and project delivery subsystem. The primary responsibilities of project analysts involve ensuring the proper flow of information throughout the project system. The classical approach to the decision process follows rigid lines of organizational charts. By contrast, the systems approach considers all the interactions necessary among the various elements of an organization in the decision process.

The various elements (or subsystems) of the organization act simultaneously in a separate but interrelated fashion to achieve a common goal. This synergism helps to expedite the decision process and to enhance the effectiveness of decisions. The supporting commitments from other subsystems of the organization serve to counterbalance the weaknesses of a given subsystem. Thus, the overall effectiveness of the system is greater than the sum of the individual results from the subsystems.

The increasing complexity of organizations and projects makes the systems approach essential to the project management environment. As the number of complex projects increases, there will be an increasing need for project management professionals who can function as systems integrators.

Project management techniques can be applied to the various stages of implementing a system as shown in the following guidelines:

1. **Systems definition:** Define the system and associated problems using keywords that signify the importance of the problem to the overall organization. Locate experts in this area who are willing to contribute to and support the effort. Prepare and announce the development plan.
2. **Personnel assignment:** The project group and the respective tasks should be announced, a qualified project manager should be appointed, and a solid line of command should be established and enforced.
3. **Project initiation:** Arrange an organizational meeting during which a general approach to the problem should be discussed. Prepare a specific development plan and arrange for the installation of needed hardware and tools.
4. **System prototype:** Develop a prototype system, test it, and learn more about the problem from the test results.
5. **Full system development:** Expand the prototype to a full system, evaluate the user interface structure, and incorporate user training facilities and documentation.
6. **System verification:** Get experts and potential users involved, ensure that the system performs as designed, and debug the system as needed.
7. **System validation:** Ensure that the system yields expected outputs. Validate the system by evaluating performance level, such as percentage of success in so many trials, measuring the level of

deviation from expected outputs, and measuring the effectiveness of the system output in solving the problem.

8. **System integration:** Implement the full system as planned, ensure the system can coexist with systems already in operation, and arrange for technology transfer to other projects.
9. **System maintenance:** Arrange for continuing maintenance of the system. Update solution procedures as new pieces of information become available. Retain responsibility for system performance or delegate to well-trained and authorized personnel.
10. **Documentation:** Prepare full documentation of the system, prepare a user's guide, and appoint a user consultant.

Systems integration permits sharing of resources. Physical equipment, concepts, information, and skills may be shared as resources. Systems integration is now a major concern of many organizations. Even some of the organizations that traditionally compete and typically shun cooperative efforts are beginning to appreciate the value of integrating their operations. For these reasons, systems integration has emerged as a major interest in business. Systems integration may involve the physical integration of technical components, objective integration of operations, conceptual integration of management processes, or a combination of any of these.

Systems integration involves the linking of components to form subsystems and the linking of subsystems to form composite systems within a single department and/or across departments. It facilitates the coordination of technical and managerial efforts to enhance organizational functions, reduce cost, save energy, improve productivity, and increase the utilization of resources. Systems integration emphasizes the identification and coordination of the interface requirements among the components in an operation. Integration can be achieved in several forms including the following:

1. **Dual-use integration:** This involves the use of a single component by separate subsystems to reduce both the initial cost and the operating cost during the project life cycle.
2. **Dynamic resource integration:** This involves integrating the resource flows of two normally separate subsystems so that the resource flow from one to or through the other minimizes the total resource requirements in a project.
3. **Restructuring of functions:** This involves the restructuring of functions and reintegration of subsystems to optimize costs when a new subsystem is introduced into the project environment.

PROJECT SYSTEMS INTEGRATION

Systems integration is particularly important when introducing new technology into an existing system. It involves coordinating new operations to coexist with existing operations. It may require the adjustment of functions to permit the sharing of resources, development of new policies to accommodate product integration or realignment of managerial responsibilities. It can affect both hardware and software components of an organization. Presented below are guidelines and important questions relevant for systems integration.

- What are the unique characteristics of each component in the integrated system?
- How do the characteristics complement one another?
- What physical interfaces exist among the components?
- What data/information interfaces exist among the components?
- What ideological differences exist among the components?
- What are the data flow requirements for the components?
- Are there similar integrated systems operating elsewhere?
- What are the reporting requirements in the integrated system?
- Are there any hierarchical restrictions on the operations of the components of the integrated system?
- What internal and external factors are expected to influence the integrated system?
- How can the performance of the integrated system be measured?
- What benefit/cost documentations are required for the integrated system?
- What is the cost of designing and implementing the integrated system?
- What are the relative priorities assigned to each component of the integrated system?
- What are the strengths of the integrated system?
- What are the weaknesses of the integrated system?
- What resources are needed to keep the integrated system operating satisfactorily?
- Which section of the organization will have primary responsibility for the operation of the integrated system?
- What are the quality specifications and requirements for the integrated systems?

The integrated approach to project management starts with a managerial analysis of the project effort. Goals and objectives are defined, a mission statement is written, and the statement of work is developed. After these, traditional project management approaches, such as the selection of an organization structure, are employed. Analytical tools including the critical path method (CPM) and the precedence diagramming method (PERT) are then mobilized. The use of optimization models is then called upon as appropriate. Some of the parameters to be optimized are cost, resource allocation, and schedule length. Some project management software has built-in capabilities for the planning and optimization needs. It should be understood that not all project parameters will be amenable to optimization.

The use of commercial project management software should start only after the managerial functions have been completed. A frequent mistake in project management is the rush to use a project management software without first completing the planning and analytical studies required by the project. Project management software should be used as a management tool, the same way a word processor is used as a writing tool. It will not be effective to start using the word processor without first organizing the thoughts about what is to be written. Project management is much more than just the project management software.

WORK BREAKDOWN STRUCTURE

Thinking in terms of finite element analysis in a digital environment, work breakdown structure (WBS) can provide the elements making up work packages in a project environment. The purpose of constructing a WBS is to analyze the elements of the project in detail. With more details, more controls are possible. Thus, WBS facilitates the itemization of the components of a project such that planning, scheduling, and control can be accomplished using available digital tools. For example, it is through the WBS elements that we may be able to determine items that can be digitally implemented alongside the items that are amenable to only analog processing. A good example in an office environment is determining what can be signed digitally compared what must be signed via wet signature. Today, almost all book contracts are signed digitally, whereby the document is routed electronically from one signatory to another. This is an expedited process compared to the traditional paper-signing requirements, whereby the document is signed and surface-mailed to all signatories in sequence.

WBS presents the inherent components of a project in a structured block diagram or interrelationship flowchart so that where digital processing is possible (or permitted) can be identified. WBS shows the relative hierarchies of parts (phases, segments, milestone, etc.) of the project. If a project is properly designed through the application of WBS at the project planning stage, it becomes easier to estimate cost and time requirements of the project. Project control is also enhanced by the ability to identify how components of the project link together. A large project may be broken down into smaller subprojects that may, in turn, be systematically broken down into task groups or work packages. Thus, WBS permits the implementation of the concept of digitally "divide and conquer" in project management.

DIGITAL INTERFACES FOR EMPLOYEE INTERACTIONS

In the digital era, employee communication is widely influenced by the use of electronic communications, such as email, Instagram, Twitter, Facebook, WhatsApp, and so on. Motivating employees and involving employees can be facilitated through these digital tools. One consideration is for how the subject line of an email can influence how people respond or not respond to emails. For example, the perpetration of the same email subject line again and again in email trails, even when the topic has veered off into new areas, can mislead employees in how they react to the email discussions. If an email is not opened, based on what is assumed to be the topic as conveyed by the subject line, then the actual and current content may not be realized. Altering the subject line of a long series of emails if the discussion has changed from what the original topic entails ensures that recipients know what is in the email.

DIGITAL MATRIX ORGANIZATION STRUCTURE

In the digital era, organizational structures may involve groups or teams that are geographically separated over wide expanses, often global. A digital matrix organization structure is more flexible in interfacing and interchanging collaborating groups within an organization. This is even more important in cases where global partners are involved. Traditionally, projects were conducted in serial functional implementations such as R&D, engineering, manufacturing, and marketing. At each stage, unique specifications and work patterns were often used without consulting the preceding and succeeding phases. The consequence was that the end product may not have possessed the original intended characteristics. For example, the first project in the series might involve the production of one component while the subsequent projects might involve the production of other components. The composite product may not achieve the desired performance because the components were not designed and produced from a unified point of view. A digital matrix organization enforces a more dynamic representation of linkages of blocks in the organizational structure. As in the traditional matrix organization, the appeal of a digital matrix organization is that it can provide synergy within groups in an organization in real time.

PROJECT PARTNERING IN THE DIGITAL ERA

Project partnering emerged in the 1990s. It involves having project team and stakeholders operating as partners in the pursuit of project goals. Benefits of project partnering include improved efficiency, cost reduction, resource sharing, better effectiveness, increased potential for innovation, and improvement of quality of products and services. Partnering creates a collective feeling of being together on the project. It fosters a positive attitude that makes it possible to appreciate and accept the views of others. It also recognizes the objectives of all parties and promotes synergism. The communication, cooperation, and coordination concepts of the new Triple C model (Badiru, 2008) facilitate partnering today. Suggestions for setting up project partnering are as follows:

- Use an inclusive organization structure
- Identify project stakeholders and clients
- Create informational linkages
- Identify the lead partner
- Collate the objectives of partners
- Use a responsibility chart to assign specific functions

PROJECT CONTROL IN A DIGITAL ENVIRONMENT

Project control requires that appropriate actions be taken to correct deviations from expected performance. Control involves measurement, evaluation, and correction. Measurement is the process of measuring the relationship between planned performance and actual performance with respect to project objectives. The variables to be measured, the measurement scales, and the measuring approaches should be clearly specified during the planning stage. Corrective actions may involve rescheduling, reallocation of resources, or expediting of tasks. In some cases, project termination is an element of project control. Project management tools can track these measurements today, but the data still must be collected and inputted. Many project leaders choose to use other existing methods. Development in this area will lead to more useful tracking tools within the software.

COMMUNICATION PATHWAY TO WORK COORDINATION

The Triple C model of communication, cooperation, and coordination (Badiru, 2008) is an effective tool for project planning and control, particularly where digital tools and techniques can be leveraged. The model can facilitate better resource management by identifying the crucial aspects of a project. The model states that project management can be enhanced by its implementation within the integrated functions of the following:

- Communication
- Cooperation
- Coordination

The model facilitates a systematic approach to project planning, organizing, scheduling, and control. It highlights what must be done and when. It also helps to identify the resources (manpower, equipment, facilities, etc.) required for each effort. Triple C points out important questions, such as:

- Does each project participant know what the objective is?
- Does each participant know his or her role in achieving the objective?
- What obstacles may prevent a participant from playing his or her role effectively?

Figure 4.2 shows a graphical representation of the Triple C model for project management in coordinated application with the DEJI model for work design. If resource management is viewed as a three-legged stool, then communication, cooperation, and coordination constitute the three legs. Communication channels provide the basis for effective communication, partnership forms the

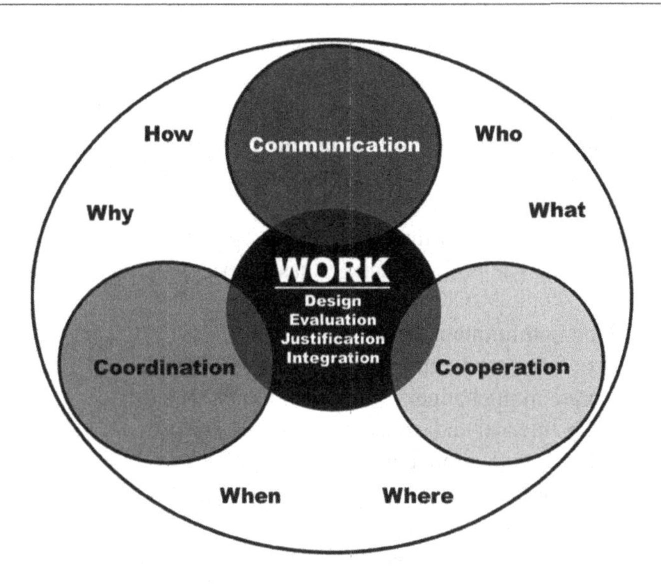

FIGURE 4.2 Triple C approach combined with DEJI model for work management.

basis for cooperation, and organizational structure provides the basis for coordination. Consequently, there must be appropriate communication channels, partnership, and proper organization structure for Triple C to be effective. This is summarized as follows:

1. For effective communication, create good communication channels.
2. For enduring cooperation, establish partnership arrangements.
3. For steady coordination, use a workable organization structure.

PROJECT COMMUNICATION

The communication function of project management involves making sure that all those concerned become aware of project requirements and progress. Those that will be affected by the project directly or indirectly, as direct participants or as beneficiaries, should be informed as appropriate regarding the following:

- The scope of the project
- The personnel contribution required
- The expected cost and merits of the project

- The project organization and implementation plan
- The potential adverse effects if the project should fail
- The alternatives, if any, for achieving the project goal
- The potential benefits (direct and indirect) of the project

Communication channels must be maintained throughout the project life cycle. In addition to internal communication, appropriate external sources should also be consulted. The project manager has several "must-do" requirements:

- Exhibit commitment to the project
- Utilize a communication responsibility matrix
- Facilitate multichannel communication interfaces
- Identify internal and external communication needs
- Resolve organizational and communication hierarchies
- Encourage both formal and informal communication links

When clear communication is maintained between management and employees and among peers, many project problems can be averted.

A communication responsibility matrix (table or spreadsheet) shows the linking of sources of communication and targets of communication. Cells within the matrix indicate the subject of the desired communication. There should be at least one filled cell in each row and each column of the matrix. This assures that each individual within a department has at least one communication source or target associated with him or her. With a communication responsibility matrix, a clear understanding of what needs to be communicated to whom can be developed.

PROJECT COOPERATION

The cooperation of the project personnel must be explicitly elicited. Merely voicing consent for a project is not enough assurance of full cooperation. The participants and beneficiaries of the project must be convinced of the merits of the project. Some of the factors that influence cooperation in a project environment include manpower requirements, resource requirements, budget limitations, past experiences, conflicting priorities, and lack of uniform organizational support. A structured approach to seeking cooperation should clarify the following:

- The cooperative efforts required
- The implications of what lack of cooperation can do to a project

- The criticality of cooperation to project success
- The organizational impact of cooperation
- The time frame involved in the project
- The rewards for good cooperation

Work used to be stated as "do what the boss/owner/manager says." The success or failure of the organization was the responsibility of the boss/owner/manager. Today, in the digital era, we have seen how management has embraced the idea of operational excellence, and the achievement of it. With the speed of communication and commerce today, everyone in the organization must be engaged in the success of the organization, or the organization fails, which will adversely affect everyone. The digital era not only created the speed and improved communications, it has also created the environment of improved cooperation. And this will only continue in the future.

REFERENCES

Badiru, A. B. (2008), *Triple C Model of Project Management: Communication, Cooperation, and Coordination*, Taylor & Francis Group/CRC Press, Boca Raton, FL.

Badiru, Adedeji B., S. Abidemi Badiru and Ibrahim A. Badiru (2008), *Industrial Project Management: Concepts, Tools, and Techniques*, Taylor & Francis Group/CRC Press, Boca Raton, FL.

Badiru, Ibrahim A. (2016), "Comments about Work Management," Interview of an Auto Industry Senior Engineer about corporate views of work design, Beavercreek, Ohio, October 29.

Hammersmith, Alan G. (2006), "Implementing a PMO—The Diplomatic Pit Bull," Workshop presentation at East Tennessee PMI Chapter Meeting, January 10.

Appendix
Glossary of Terminologies in the Digital Era

Understanding the language of the digital era can go a long way in understanding the potentials and pitfalls in the digital operations and transactions. This appendix presents a collection of commonly encountered terminologies in the digital era. The glossary is culled from several sources including Badiru et al. (2017), Badiru and Racz (2016), Badiru (2014), Badiru and Thomas (2009), and Kerrigan (2021). We can learn a lot about the digital era by understanding the terminologies that represent the era. While this appendix is not all-inclusive, it does provide a basic introduction to the diverse terminologies that a prevalent in the digital era. News terms evolve frequently, particularly as new digital platforms are developed.

5G (Fifth-Generation): 5G is the latest generation of mobile technology, which is an advancement beyond the present 4G technology. It has enhanced capabilities that provide the platform for new and emerging ICT enterprises, such as Internet of Things (IoT), Artificial Intelligence (AI), Virtual Reality (VR), and Augmented Simulation. These developments will improve the way we live, work, and interact as human beings.

Accessibility: Accessibility is the process of making web content accessible to people with disabilities such as those who are visually impaired, hearing-impaired, color blind people, or anyone else who cannot for whatever reason, use a computer in a conventional manner. A website with poor accessibility will be difficult for these people to use. Accessibility is particularly important for sites providing information to those with disabilities such as those in the health care sector and government departments. Accessibility is an important aspect to consider when designing any site. Guidelines for the various levels of accessibility are set by industry bodies (Kerrigan, 2021).

AJAX: Stands for Asynchronous JavaScript and XML. AJAX is typically used for creating dynamic web applications and allows for asynchronous data retrieval—content updating without the web page having to reload. The JavaScript on any given page handles most of the basic functions of the application, making it perform more like a desktop program instead of a web-based one. A good example of AJAX in action is Google Maps—which allows the visitor to scroll around a map without the browser refreshing.

Algorithm: A set of instructions or procedures used in order to accomplish a task, such as creating search results on the Internet (Badiru et al., 2017). In the context of search, algorithms are used to provide the most relevant results first based on those instructions.

AMA: Ask me anything. A style of Reddit post, in which the poster opens themselves up to questions and answers them in the comments, usually done in real time.

Analytics: Measuring, analyzing, and interpreting activity and interactions on and across social media platforms (Badiru and Thomas, 2009). It is a powerful tool for marketers, enabling them to see how they are performing against Social Media Objectives and where they can improve.

Anchor text: Anchor text usually gives the user relevant descriptive or contextual information about the content of the link's destination. The anchor text above in bold—what the visitor will see on a website.

Android: Usually used in the context of Android phone, Android is a free and open-source operating system developed by Google that powers a variety of mobile phones from different manufacturers and carriers. It is a rival of the iPhone platform. In contrast to Apple's tightly controlled architecture and App Store, Android allows users to install apps from the Android Market and from other channels, such as directly from a developer's website.

API (application programming interface): The way computer programs share data and functionality with other computer programs. APIs are an increasingly critical part of the Internet's interconnection. Many say that the future of the Internet lies in APIs because they help distribute and combine content. On the Web, APIs are generally special URLs that give back machine-readable data, in formats like JSON or XML, rather than human-readable data, which is usually HTML. Facebook, Twitter, and Google Maps all have APIs that allow other websites or computer programs to use their underlying tools. *The New York Times* and National Public Radio (NPR) have also released APIs that allow other programs to draw on archives of movie reviews, restaurant reviews, and articles.

App: Short for application, a program that runs inside another service. Many mobile phones allow apps to be downloaded, leading to a burgeoning economy for modestly priced software. Can also refer to a program or tool that can be used within a website. Apps generally are built using software toolkits provided by the underlying service, whether it is iPhone or Facebook.

ASP: Active Server Pages (ASP), also known as Classic ASP or ASP Classic, was an early server-side script engine for dynamically generated web pages.

ATL: Above the line ads include any which focus on general media such as TV, cinema, radio, print, and the Internet.

Atom: A syndication format for machine-readable web feeds that is usually accessible via a URL. While it was created as an alternative to RSS (Real Simple Syndication) to improve upon RSS's deficiencies (such as ambiguities), it still is secondary to RSS. (See also, RSS)

Audience segment: A group of users with similar traits or characteristics.

Audience targeting: Targeting of specific audience segments, such as an age demographic. Audience segments can be defined in OpenX or in an external data management platform (DMP).

Audit trail: Logging of any changes to data (creation, modification, or deletion) to allow a system admin user to review all historical changes, particularly when something disastrous or undesirable has occurred (Badiru and Racz, 2016).

Avatar: A picture or a graphic that represents you or your company online. Also commonly known as a profile picture. For example, in Whatsapp social media app, dp (display photo) is used.

Back end: The back end of a website is the part hidden from view of regular website visitors. The back end generally includes the information structure, applications, and the content management system (CMS) controlling content on the site.

Backlink: Backlinks are links from other websites back to yours. They are sometimes also referred to as "trackbacks" (especially on blogs). Backlinks have a huge impact on your sites search rankings. Lots of backlinks from high-ranking sites can greatly improve search engine results, especially if those links use keywords in their anchor text.

Bad neighborhood server: A "bad neighborhood" refers to the server where your site is hosted. A site hosted on a server that hosts other sites that spam or use black hat SEO practices can end up penalized by search engines solely because of their proximity to those sites. Linking to sites in bad neighborhoods can also have a negative effect on search rankings.

Bandwidth: Bandwidth can refer to two different things: the rate at which data can be transferred, or the total amount of data allowed to be transferred from a web host during a given month (or other hosting service term) before overage charges are applied. It is generally referred to in term of bits per second (bps), kilobits per second (kbs), or other metric measurements. Lower bandwidth Internet connections such as dial-up mean data load slower than with high-bandwidth connections such as broadband.

Banner: This is an ad that appears on a web page, which is typically hyperlinked to an advertiser's website. Banners can be images (GIF, JPEG, PNG), JavaScript programs, or multimedia objects (Flash, Java, Shockwave, etc.).

Below the fold: This term is a carryover from newspaper publishing days. In newspaper terms, "below the fold" means content was on the bottom half of the page below the physical fold in the paper. In web design terms, "below the fold" refers to the content that is going to appear off of the bottom of the screen for the average website visitor in their browser.

Blog: A personal or company website or web page on which an individual or group of individuals post informal articles, record opinions, or share links, videos, or imagery, on a regular basis.

Blogger outreach: The process by which you "reach out" to other bloggers within your industry or category. Blogger outreach involves contacting individual bloggers and asking them if they are interested in either hosting content you've created for them or in writing content and mentioning or linking to your website or brand.

Bloggersphere or blogosphere: A term to refer to all known blogs on the Internet and their interconnections.

Bounce rate: Website bounce rate is the number, or percentage of people who leave the site from the same page they entered, without clicking through to any other pages. This can be a good indicator of how good the navigation on the website is, as well as an indicator of the quality of the content on the website. Generally, a high bounce rate indicates a poor page design.

Breadcrumb: Breadcrumbs are the navigational elements that generally appear near the top of a given web page that show you the pages and subpages that appear before the page you're on. For example, on a blog, the breadcrumb might look something like: Home > Category > Year > Month > Post, or maybe a lot simpler. The term "breadcrumb" comes from the fairy tale "Hansel and Gretel."

Browser: Browser refers to the program a website visitor is using to view the website. Examples include Safari, Firefox, Google Chrome, Opera, and Internet Explorer.

Cache/Caching: Cached files are those that are saved or copied (downloaded) by a web browser so that the next time that user visits the site, the page loads faster.

Cascading Style Sheets (CSS): Also referred to simply as CSS, Cascading Style Sheets are used to define the look and feel of a website outside of the actual HTML of the site. In recent years, CSS has replaced tables and other HTML-based methods for formatting and laying out websites. The benefits of using CSS are many, but some of the most important are the simplification of a site's HTML files, which can actually increase search engine rankings and the ability to completely change the style of a site by changing just one file without having to make changes to content. The latest version is CSS3.

CGI: The Common Gateway Interface (CGI) is a standard protocol for interfacing external application software with an information server, commonly a web server. CGI usually comes in the form of a script, which is called to perform a particular function on a website. CGI scripts are usually written in a programming language called PERL.

Civic media: An umbrella term describing media technologies that create a strong sense of engagement among residents through news and information. It is often used as a contrast to "citizen journalism" because it also encompasses mapping, wikis, and databases.

Client-side: Client-side refers to scripts that run in a visitor's browser, instead of on a web server—as in server-side scripts. Client-side scripts are generally faster to interact with, though they can take longer to load initially.

Cloud computing: An increasingly popular computing model in which information and software are provided on demand from over the Internet rather than staying on local computers. Cloud computing is appealing because companies can reduce the amount they spend on their own computer servers and software but can also quickly and easily expand as the company grows. Examples of cloud computing applications include Google Docs and Yahoo Mail. Amazon offers two cloud computing services: EC2, which many start-ups now use as a cheap way to launch their products, and S3, an online storage system many companies use for cheap storage.

Comment: In web design terms, a comment is a bit of information contained in the HTML or XHTML of a web page, which is ignored by the browser. Comments are used to identify different parts of the file and as reference notes. Good commenting makes it much easier for a web designer to make changes to the site, as it clearly defines which parts of the code perform what functions. There are different comment formats for different programming and markup languages.

Community manager: Responsible for building, growing, and managing online communities around a brand, cause, or topic.

Content management system (CMS): Also known as a CMS, the content management system is a backend tool for managing the content of a website. The CMS separates the core content (text, images, video) from the design and functionality of the site. Using a CMS generally, makes it easier to change the design or function of a site independently of the website content. It will often make it easier for administrators who aren't web developers, to add content to a website.

Content marketing: A marketing technique of creating and distributing relevant and valuable online content to attract, acquire, and engage a clearly defined target audience.

Content strategy: The planning and development for the creation, delivery, and governance of online, shareable content. A content strategy should outline what social channels to use, what types of content are to be shared via what channels, and a schedule of content production.

Conversion: A conversion is the completion of a goal by an end user of a website—having gone through a series of steps to reach that goal. A goal can be defined as anything, ranging from the user completing a registration form or survey, downloading a file, or purchasing something. Conversion rate is a good measure of the performance of a website and will often be the benchmark used when trying to improve the efficiency of a website.

CPA (cost per action): A pricing model in which the advertiser is charged for an ad based on how many users take a specific, pre-defined action—such as buying a product from an online store—based on viewing an ad. This is the "gold standard" for advertisers because it most directly matches the cost of an ad to its effectiveness. However, it's not commonly used since it's extremely difficult to measure: it is often unclear when or how to attribute an action to a specific ad. (Also sometimes referred to as cost per acquisition.)

CPC (cost per click): A pricing model in which the advertiser is charged for an ad based on how many users click it. This is a common model for "search advertising" (the all-text ads associated with search results) and with text ads in general. CPC is well suited for "directed" advertising, intended to prompt an immediate response because a user's clicking on an ad shows engagement with it. Google AdWords is generally priced on a CPC basis.

CPM (cost per mille): Cost per one thousand (often views). Much of online advertising—particularly display advertising—is priced on a CPM basis. (Mille = Latin for one thousand; we use "K" for "kilo" almost everywhere else in tech, but "M" for "mille" here, which causes some confusion). CPM is well suited for "brand" or "awareness" advertising, in which the primary purpose of the ad is not necessarily to prompt an immediate response.

CPM (critical path method): Project scheduling techniques using a network of nodes and arrows to represent activities and their precedence interrelationships.

Creative Commons: A flexible set of copyright licenses that allow content creators to specify which rights they reserve and which they waive regarding their work that is supposed to codify collaborative spirit of the Internet. There are six main Creative Commons licenses based on four conditions that creators can choose to apply: Attribution, Share Alike, Non-Commercial, and No Derivative Works. The least restrictive of the licenses is Attribution, which grants anyone, from an individual to a large company, the right to distribute, display, or otherwise make use of the work so long as the creator is credited.

CSS: See Cascading Style Sheets.

CSS framework: A CSS framework is a collection of CSS files used as the starting point to make XHTML and CSS websites quickly and painlessly. They usually contain CSS styles for typography and layout.

CSV (comma-separated values): An extremely simple data format that stores information in a text file. CSV is popular precisely because it can be easily read by many different applications, including spreadsheets, word processors, programming text editors, and web browsers. Thus it is a common way for people, including governments, to make their data available. Each row of data is represented by a line of text. Each column is delimited/separated by a comma (,). To prevent confusion about commas in the data, the terms are often surrounded by double quotes (").

Data visualization: A growing area of content creation in which information is represented graphically and often interactively. This can be used for subjects as diverse as an analysis of a speech by the president and the popularity of baby names over time. While it has deep roots in academia, data visualization has begun to emerge on content sites as a way to handle the masses of data that are being made public, often by government.

DEJI systems model: A system engineering methodology for system Design, Evaluation, Justification, and Integration. See www.dejimodel.com.

Deprecated: Deprecated code is code that is no longer included in the language specifications. Generally, this happens because it is replaced with more accessible or efficient alternatives. This happens when a new version of a programming language or framework is released.

DHTML: Stands for dynamic hypertext markup language. DHTML fuses XHTML (or any other markup language), the DOM, JavaScript (or other scripts), and CSS (or other presentation definition languages) to create interactive web content.

Django: A web framework that is popular among news and information sites, in part due to its origin at *Lawrence Journal-World* in Kansas. It is written in Python, a sophisticated dynamic language.

DNS: Stands for "Domain Name Service" or alternately "Domain Name System," or "Domain Name Server." The DNS converts IP addresses into domain names. DNS servers are provided with the IP address of your web server when you assign your domain name to those servers. In turn, when someone types your domain name into their web browser, the DNS servers translate the domain name to the IP address and point the browser to the correct web server.

Doctype: The doctype declaration specifies which version of HTML is used in a document. It has a direct effect on whether your HTML will validate or not.

Document-oriented database: An increasingly popular type of database. In contrast to relational databases, which rigidly require information to be stored in pre-defined tables, document-oriented databases are more free-flowing and flexible. This is important when you don't know what is going to be thrown at you. Document-oriented databases retrieve information more quickly but store it less efficiently. The same document-oriented database might let you store the information for an article (headline, byline, data, content, miscellaneous) or for a photo (file, photographer, date, cutline).

DOM: Stands for Document Object Model. It is a language-independent, cross-platform convention for representing objects in XML, XHTML, and HTML documents. Rules for interacting with and programming the DOM are specified in the DOM API.

Domain: The domain is the name by which a website is identified. The domain is associated with an IP address. Domains can be purchased with any combination of letters, hyphens (-), and numbers (though it can't start with a hyphen). Depending on the extension (.com, .net, .org, etc.), a domain can be anywhere up to 26–63 characters long.

Domain authority: This is a scale from 1 to 100 that search engines use to determine how authoritative a company's website is, 1 being the lowest rank and 100 being the highest. The higher the domain authority, the more search engines trust the domain owner.

Drupal: A popular **content** management system known for a vibrant open-source community that creates diverse and robust extensions. Drupal is very powerful, but it is somewhat difficult to use for simple tasks when compared to WordPress. Drupal provides options to create a static website, a multiuser blog, an Internet forum, or a community website for user-generated content. It is written in PHP and distributed under the General Public License (GPL) open-source license.

DTD: Stands for Document Type Definition. DTD is one of several SGML and XML schema languages. It provides a list of the attributes, comments, elements, entities, and notes in a document along with their relationships to each other.

Earned media: Word of mouth mentions of your company, generated on a person-to-person basis EC2. A computing power rental system by Amazon that has become popular among technology companies because it is much cheaper than maintaining your own computer servers. Users can host their applications on EC2 and pay depending on usage. EC2 is an example of cloud computing. (Also see cloud computing)

E-commerce: Short for electronic commerce, e-commerce is the process of buying and selling of goods online through websites. Products sold through e-commerce can be physical products that require shipping, or digital products delivered electronically such as program and document downloads, license keys, and music.

Elastic layout: An elastic layout is one that uses percentage and em for defining page width, paired with a max width style to allow the layout of the website to stretch when font sizes are changed. The ability to flex to accommodate the browser width and visitor's font preferences is where the name "elastic" comes from.

Element: In XML, an element is the central building block of any document. Individual elements can contain text, other elements, or both.

Em: Em is a unit of measurement for sizing fonts and other elements within a web page relative to the parent element of the item. A 1em font is equal to the point size for the font already defined in the parent element. For example, 2em would be twice the current size, .5em would be half the current size, and so on.

Embedded style: An embedded style is a CSS style written into the head of an XHTML document. It only affects the elements on that page, instead of site-wide as a separate CSS file does. Style in an embedded style sheet will override styles from the linked CSS file.

Engagement: Engagement or social engagement refers to the amount of participation in an online community. Retweets, comments, and likes are all forms of social engagement.

Event tracking: The code on web pages that allows Google Analytics to track and measure events on web pages, such as clicks or downloads.

Ex: Ex is a measurement for font height or size relative to the height of a lowercase "x" in that font family.

Extensible Markup Language: Otherwise known as XML. XML is a markup language used for writing custom markup languages. In other words, XML describes how to write new languages. It is sometimes referred to as a "meta" language because of this. It also serves as a basic syntax that allows different kinds of computers and applications to share information without having to go through multiple conversion layers.

External style sheet: This is CSS that is stored in an external document. The biggest advantage to using an external style sheet is that it can be linked to by multiple HTML/XHTML files, meaning that changes made to the style sheet will affect all the pages linked to it, rather than having to change each page individually.

Facebook: Facebook is a social utility that connects people with friends and others who work, study, and live around them. People use Facebook to connect and share content with friends and associates. Facebook is now the largest social network on the Internet, with more than 400 million active users, half of whom log on to Facebook in any given day. People spend over 500 billion minutes per month on Facebook.

Facebook Connect: A technology from Facebook that allows a reader to log into a third-party website with their Facebook account, rather than creating a new profile for that website. Facebook Connect, which is an API, also allows the third parties to pull certain data from the user's profile, such as his or her name and age. In turn, the reader's activities on the website can also be displayed on her or his Facebook profile.

Facebook fan page: A Facebook profile for a specific person, product, company, or organization, usually administered by official representatives. This is different from a Facebook personal page, which must be owned by an individual, and different from a Facebook community page, which is built around an interest not related to a brand, such as "cooking." It is also different from a Facebook group. Fan pages can gather thousands or millions of fans through "likes," and official posts by the page administrator generally go into the fans' news streams. Once a page has more than 25 fans, it can claim a short form URL, such as facebook.com/nytimes or facebook.com/wikileaks. Facebook community and fan pages are strong players in ongoing efforts to bring content to people where they already are, instead of requiring them to come to the content.

Facebook group: Facebook groups are analogous to offline clubs. Unlike Facebook fan pages, groups do not have to be administered by official representatives. In addition, the activity posted in groups does not get pushed into users' feeds. But as long as it has fewer than 5,000 members, Facebook groups are allowed to mass message all their members.

Facebook personal page: A profile page tied to a single individual. What information is controlled (in theory) by the individual. However, because there is a 5,000-person limit to friends, some celebrities have fan pages instead. As of 2009, individuals can choose a username, which makes their page available at facebook.com/username.

Facetime: A free video chat and VOIP software application developed by Apple to run on all devices running iOS, the Apple operating system found on iPhones, iPads, and computers.

Favicon: Favicons are tiny (generally 16 × 16 pixels, though some are 32 × 32 pixels), customizable icons displayed in the web address bar in most browsers next to the web address. They're either 8-bit or 24-bit in color depth and are saved in either .ico, .gif, or .png file formats.

Fixed width layout: A fixed width web page layout has a set width, usually defined in pixels. The width stays the same regardless of screen resolution, monitor size, or browser window size. It allows for minute adjustments to be made to a design that will stay consistent across browsers. Designers have more control over exactly how a site will appear across platforms with this type of layout.

Flash: A proprietary platform owned by Adobe Systems that allows for drag-and-drop animations, program interactivity, and dynamic displays for the Web. The language used, ActionScript, is owned by Adobe; this contrasts with many other popular programming languages that are open source. Creators must use Adobe's Creative Suite products and web surfers must install a Flash plug-in for their browser.

Fluid layout: See Liquid Layout.

Flickr: An image and video hosting website popular with bloggers and owned by Yahoo! Forums—An online discussion site where registered users can hold conversations in the form of posted messages, or threads.

Focal point: The focal point of a website is the spot on a web page that the eye is naturally drawn to. This could be an image, a banner, text, Flash content, or just about anything else. It is important to ensure that the focal point is the most important part of the page.

Font family: A font family is a group designation for defining the typefaces used in CSS documents. The font family tag generally lists multiple fonts to be used, and usually ends with the generic font category such as "serif" or "sans-serif."

Font style: In CSS, the font style refers solely to whether a font is italic or not.

Font weight: The font weight refers to how thick or thin (bold or light) a font looks.

Foursquare: Foursquare is a geo-social network where users can "check-in" to places they are visiting, leave tips, add reviews, and post pictures. Geo-social works by combining a traditional social networking model of registered users who connect with one another, through Global Positioning Satellite (GPS) technology. By connecting these two models, users don't just get updates, they also get information about where the updates were made.

Framework: A software package that makes writing programs easier by providing all the "plumbing" for a particular type of task (like writing a web app), allowing programmers to just "fill in the blanks" with their own project-specific needs.

Front-end: The front-end is the opposite of the back-end. It is all the components of a website that a visitor to the site can physically see such as pages, images, and other content. More specifically, it is the interface that visitors use to access the website content. It is also sometimes referred to as the user interface (UI).

Geolocation/Geotagging: Geolocation is the process of finding, determining, and providing the exact location of a computer or networking device. It enables the use of a variety of location-based apps, such as Foursquare.

Goal tracking: The code on web pages that allows Google Analytics to track when certain tasks are completed, such as a download.

Google: The World's largest Internet search engine, indexing literally billions of web pages. In May 2010, 30.7 m people in the United Kingdom searched on Google.

Google AdSense: Google's online advertising network that allows content publishers to embed a piece of code to display Google ads on their sites. The ads are selected based on the content of the page. Ad revenue is split between Google and the publisher in an undisclosed proportion, generally believed to be two-thirds to the publisher. (Note: ads on Google's own sites are covered by Google AdWords, not AdSense.)

Google AdWords: Google's text-based flagship advertising product, which provides the lion's share of the company revenue. Ads are displayed on Google's own sites based on search terms that users type in, and advertisers pay only when the users click on them. The search terms, called keywords, are purchased by advertisers; availability of a given keyword is based in part on an auction system, and in part on the responsiveness of the audience.

Google Analytics: A Google tool used to measure data about your website's performance.

Google Authorship: A way of linking the content you create, on any website, with your Google+ profile. When you have correctly set up Google Authorship, your profile picture will appear in Google search results beside any written content you publish online. As Google's search algorithm favors content from thought leaders and experts, using Google Authorship is a way to boost your ranking.

Google Hangouts: An instant messaging and video chat platform developed by Google that works across all computers and both Android and Apple phones and tablets.

Google+: A social network owned and operated by Google. Users can post stories, share videos and pictures, organize friends into circles, join communities, and chat to other users. Your Google+ account syncs up with all other Google accounts you may have; Gmail, YouTube, etc.

Graceful degradation: Graceful degradation refers to the ability for a website to have elements that may take advantage of the capabilities of newer browsers, but done in a way that allows users with older browsers to still view the site in a manner that at least allows them access to basic content.

Graphical user interface: Also referred to by its acronym: GUI. A graphical user interface uses an input device such as a mouse and gives visual representations of how the user is able to interact with a web application. In other words, a GUI is all of the front-end stuff you see on a web application. The purpose of a GUI is to allow interaction with a web application without having to enter code.

Hashtag: A word or unspaced phrase prefixed with the # symbol. Commonly used on social networking sites such as Twitter, Google+, and Instagram. Appending a hashtag to a post both helps to give that message context and groups it with other messages about that same topic or event.

Hexadecimal: Also referred to a "hex" numbers, they are a base-16 numbering system used to define colors online. Hex numbers include the numerals 0–9 and letters A–F. Hexadecimal numbers are written in three sets of hex pairs. Because screen colors are Red, Green, and Blue, the first pair defines the Red hue, the second pair defines the Green hue, and the third pair defines the Blue.

Hit: Contrary to popular belief, a hit does not represent a single visitor to a website. A hit is actually a request for a single file from your web server. This means one page can actually generate multiple hits, as each page generally has more than one file, for example, a HTML or other base file, a CSS file, multiple images, and so on. Each request for each file is one hit, so a web page with 50 images on would register 50 hits.

Hootsuite: A social media management platform that allows users and brand managers to update, post, schedule, and respond via multiple social media platforms, including for Facebook, Twitter, LinkedIn, Google+, and WordPress.

htaccess: The .htaccess file is the default directory-level configuration file on Apache web servers. It is also known as a "distributed configuration file."

HTML: (Hypertext Markup Language)—The dominant formatting language used on the World Wide Web to publish text, images, and other elements.

HTML tag: Also referred to as an HTML element, an HTML tag is the bit of code that describes how that particular piece of the web page it is on is formatted. Typical tags specify things like headings, paragraphs, links, and a variety of other items.

HTML5: The upcoming, powerful standard of Hypertext Markup Language, which has added advanced interactive features, such as allowing video to be embedded on a web page.

HTTP: Stands for Hypertext Transfer Protocol. HTTP is a set of rules for transferring hypertext requests between a web browser and a web server.

HTTPS: Similar to HTTP, HTTPS stands for Hypertext Transfer Protocol over SSL (Secure Socket Layer) or, alternately, Hypertext Transfer Protocol Secure. Like HTTP, it is a set of rules for transferring hypertext requests between browsers and servers, but this time it is done over a secure encrypted connection.

Hyperlink: A hyperlink is a link from one web page to another, either on the same site or another one. Generally, these are text or images, and are highlighted in some way; text is often underlined or put in a different color or font weight. The inclusion of hyperlinks creates the "hyper" part of "hypertext."

Hypertext: Hypertext is any computer-based text that includes hyperlinks. Hypertext can also include presentation devices like tables or images, in addition to plain text and links.

Iframe: An HTML tag that allows for one web page to be wholly included inside another; it is a popular way to create embeddable interactive features. Iframes are usually constructed via JavaScript as a way around web browsers' security features, which try to prevent JavaScript on one page from quickly talking to JavaScript on an external page. Many security breaches have been designed using iframes.

Image Map: An image map is used in XHTML to allow different parts of an image to become different clickable elements. It can also allow some portions of an image to have no clickable element.

Infographic: An infographic (information graphic) is a graphical representation of information or data designed to be engaging and easily understandable.

Inheritance: In CSS, elements that do not have a pre-defined style will take on the style of their parent element within the document tree.

Inline style: Elements with CSS written directly around the element it affects, instead of in a separate style sheet or header style.

Instagram: A photo- and video-sharing social network that enables registered users to take pictures and videos via the Instagram mobile app, apply digital filters to them, and then share them across other social media profiles, such as Facebook, Twitter, Tumblr, and Flickr.

Integrated media: When multiple of the big three media types (paid, owned, and earned) overlap and work together in a marketing campaign.

Intelligent display: Using data and consumer behavior to generate and serve a custom advertisement to a specific person.

IP Address (Internet Protocol Address): A unique string of numbers that identifies each device connecting to the Internet. The IP Address is your website's home address.

JavaScript: A Web scripting language used to enhance websites; it can make them more interactive without requiring a browser plug-in. JavaScript is interpreted by your browser instead of by a web server, otherwise known as a client-side scripting language. JavaScript files generally end in .js. Despite its name, it is not related to the Java language.

Joomla: A free, open-source content management built in PHP. It is more powerful than WordPress but not as powerful as Drupal. However, it is known for its extensive design options. The name Joomla means "all together" in Swahili.

jQuery: An incredibly popular open-source JavaScript library designed for manipulating HTML pages and handling events. Released in 2006, jQuery quickly gained widespread adoption because of its efficiency and elegance. The definitive feature of jQuery is its support for "chaining" operations together to simplify otherwise complicated tasks.

JSON (JavaScript Object Notation): A Web data publishing format that is designed to be both easily human and machine-readable. It is an alternative to XML that is more concise because, unlike XML, it is not a markup language that requires open and close tags.

Key/value store: A simpler way of storing data than a relational or document database. Key-value stores have a simple structure, matching values to accessible "keys," or indices. In Web development, key/value stores are often (though not always) used for optimization.

Keyword: A keyword is a word or phrase that your audience uses to search for relevant topics on search engines.

Keyword stuffing: This is the practice of using too many keywords in content in hopes of making it more visible on search engines. You will be penalized by search engines if you resort to it. Never keyword stuff, just provide great and valuable content.

Klout: A website and mobile app that uses social media analytics to rank users according to online social influence. When calculating a person's score, Klout takes into account their activity across several social media platforms, including Twitter Facebook, LinkedIn, Google+, and Instagram.

LAMP: Stands for Linux, Apache, MySQL, and PHP (or sometimes Perl or Python), and is referring to the specifications of a web server (defining the operating system, web server, database, and scripting language, in that order).

Landing page: A landing page is the page where a visitor first enters a website. Often, a special landing page is created to elicit a specific action from the new visitor—usually in connection with an advertising or marketing campaign.

Legacy media: An umbrella term to describe the centralized media institutions that were dominant during the second half of the 20th century, including—but not limited to—television, radio, newspapers, and magazines, all which generally had a unidirectional distribution model.

Library: In the context of programming, this contains code that can be accessed for software and Web development, enabling one to perform common tasks without writing new code every time. Many libraries are freely shared.

"Like": Clicking "Like" under something someone posts on Facebook lets them know that you enjoy it, without leaving a comment. Just like a comment though, the fact that you liked it is noted beneath the item.

Link farm: A link farm is any website set up specifically to increase the link popularity of other websites by increasing the number of incoming links to that site. While some link farms are single pages listing unrelated links, others consist of networks of sites that contain multiple links back and forth to one another. Search engines can generally recognize these types of schemes and often remove link farms from their directories and penalize the sites linking to and from them.

LinkedIn: A social networking site for professionals. Registered users may list their employment and educational history on their profiles, connect with friends and colleagues, follow companies, join interest groups, share content, and have discussions.

Liquid layout: A liquid layout is one that is based on percentages of the browser window size. The layout of the site will change the width of the browser, even if the visitor changes their browser size while viewing the page. Liquid layouts take full advantage of the browser width a visitor is using, optimizing the amount of content you can fit onscreen at one time.

Live blogging: Similar to live television or live radio, live blogging is a blog post or series of posts intended to provide real-time textual coverage of an event.

Location-based services: A service, usually in a mobile Web or mobile device application, that uses your location in order to perform a certain task, such as finding nearby restaurants, giving you directions, or locating your friends.

Markup: This refers to the coding applied to a text document to change it into an HTML, XML, or other markup language document.

Mashup: A combination of data from multiple sources, usually through the use of APIs. An example of a mashup would be an app that shows the locations of all the movie theaters in a particular town on a Google map. It is mashing up one data source (the addresses of movie theaters) with another data source (the geographic location of those addresses on a map).

Meetup: The world's largest network of local groups. It enables users to organize a local group or find one of the thousands already meeting up face-to-face.

Meme: An image, video, piece of text, or any combination of the three, typically humorous in nature, that is copied and spread rapidly by users across the Internet, often starting on sites such as Reddit, and often with slight variations.

Meta tag: A Meta tag is an HTML tag used to include Metadata within the header of a web page.

Metadata: Data about data. Examples of metadata include descriptors indicating when information was created, by whom and in what format. Metadata helps to organize information online and make it machine-readable. HTML is an example of metadata—it organizes the data in a web page so browsers can display it sensibly. Web pages often have hidden metadata that helps with their search engine ranks.

Micropost or Microblog: A microblog is similar to a traditional blog in every way except its size. Microblogs are much smaller, sometimes comprising of nothing more than one sentence. Microposts have been made popular by social networking site Twitter, as its users are limited to posts of no more than 140 characters in length.

MySpace: A social networking site and mobile app with a strong emphasis on music. It was the most visited social networking site in the world from 2005 until 2008 when it was overtaken by Facebook.

Navigation: Navigation refers to the system that allows visitors to a website to move around that site. Navigation is most often thought of in terms of menus, but links within pages, breadcrumbs, related links, pagination, and any other links that allow a visitor to move from one page to another are included in navigation.

Nesting: Nesting refers to putting one HTML element within another element. When this is done, the elements have to be closed in the reverse order from how they were opened.

.NET: The .NET Framework is Microsoft's comprehensive and consistent programming model for building applications.

Newsfeed: The homepage of most social networking sites, a news feed is a constantly updating list of stories from people and companies that a person follows on that social network.

Ning: An online platform for people, businesses, and organizations to create custom social networks each with customizable appearance feel and features.

Non-breaking space: A non-breaking space, also referred to as is a white-space character that is not condensed by HTML. The primary function of a non-breaking space is to hold open table cells or add spacing between words, or the beginning of paragraphs if an indent is desired.

OAuth: A new method that allows users to share information stored on one site with another site. For example, some web-based Twitter clients will use OAuth to connect to your account, instead of requiring you to provide your password directly to that third-party site.

Online community: A virtual community consisting of people and/or businesses with common interests who use social networking sites, instant messaging apps, and online forums to communicate and work together.

Ontology: A classification system with nodes or entities, that allows nonhierarchical relationships, in contrast to a taxonomy, which is hierarchical. Taxonomies and ontologies are important in content to help related articles or topics pages. (Also see taxonomy)

OOP: Object-oriented programming (OOP) is a programming paradigm that uses "objects"—data structures consisting of data fields and methods together with their interactions—to design applications and computer programs. Programming techniques may include features such as data abstraction, encapsulation, modularity, polymorphism, and inheritance.

Open Graph: Web code that determines what information (title, image, description, etc.) is displayed when someone shares a link from your website onto a social network.

Open ID: An open standard that lets users log in to multiple websites using the same identity through a third party. It is supported by numerous sites, including LiveJournal, Yahoo!, and WordPress.

Open source: Open source refers to the source code of a computer program made available to the general public. Open-source software includes both web-based and desktop applications.

Open source: Open source refers to a philosophy and a means of developing and licensing software and other copyrighted works so that others are free to inspect, use, and adapt the original source material. There are many open-source licenses.

Operating system: A basic layer of software that controls computer hardware, allowing other applications to be built on it. The most popular operating systems today for desktop computers are the various versions of Microsoft Windows, Mac OS X, and the open-source Linux. Smartphones also have operating systems. The Palm Pre uses webOS, numerous phones use Google's Android operating system, and the iPhone uses iOS (formerly known as iPhone OS).

Opt-in: The process by which a user agrees to receive messages from a company. Opt-in messages, therefore, cannot be considered as spam.

Opt-out: The process by which a user elects to stop receiving messages from a company. If a user continues to receive messages after opting out, these messages can be considered to be spam.

Organic traffic: This is traffic that is generated to your website, which is generated by a search engine. This could be traffic from Google, Yahoo, or Bing. It's also known as "Free" traffic. Organic traffic is the best type of traffic!

Owned media: Everything your company has direct control over the messaging on, including your website, your brochures, your call intake staff, and your social media channels.

Page view: A page view is a request for an entire web page document from a server by a visitor's browser. In other words, for each page view your site had, someone (or a search engine spider) looked at that page.

Paid traffic: Paid search is when a company bids on keywords and makes advertisements around those keywords to be displayed on search engines. These results appear separately, either on the top, bottom, or right side of a search results page. Paid traffic also encompasses any form of paid advertisement that directly points to your website.

Perl: Perl is a high-level, general-purpose, interpreted, dynamic programming language. Perl was originally developed by Larry Wall, a linguist working as a systems administrator for NASA, in 1987, as a general-purpose UNIX scripting language to make report processing easier. Peer-to-peer (P2P)—A network architecture in which users share resources on their own computers directly with others.

Permalink: Short for "permanent link." Generally used only on blogs, a permalink is a link that is the permanent web address of a given blog post. Since most blogs have constantly changing content, the permalink offers a way for readers to bookmark or link to specific posts even after those posts have moved off the home page or primary category page.

PHP: A recursive acronym for Hypertext Preprocessor, PHP is a widely used general-purpose scripting language that is especially suited for Web development and can be embedded into HTML.

Pinterest: A pinboard-style photo-sharing website that allows registered users to create and manage theme-based image collections (called "boards") such as events, interests, and hobbies. Users can "pin" content found around the web to boards they've created.

Pixel: A contraction of picture element, a pixel refers to a single point in a graphic. Ad units are typically measured in pixels, for example, the default 468 × 60-sized banner.

Platform: In the technology world, platform refers to the hardware or software that other applications are built upon. Computing platforms include Windows PC and Macintosh. Mobile platforms include Android, iPhone, and Palm's WebOS.

Plug-in: A plug-in is a piece of third-party code that extends the capabilities of a website. It is most often used in conjunction with a CMS or blogging platform. Plug-ins are a way to extend the functionality of a website without having to re-write the core code of a website. Plug-ins can also refer to pieces of third-party software installed within a computer program to increase its functionality.

Posterous: A blogging and publishing platform to which users can submit via email. Through APIs, it can push the content to other sites such as Flickr, Twitter, and YouTube.

PostgreSQL: An alternative to MySQL, another free and open-source relational database management system on the Internet. PostgreSQL is preferred by some in the technology community for its ability to operate as a spatial database, using PostGIS extensions.

Programming language: A special type of language used to unambiguously instruct a computer how to perform tasks. Programming languages are used by software developers to create applications, including those for the web, for mobile phones, and for desktop operating systems. C, C++, Objective C, Java, JavaScript, Perl, PHP, Python, and Ruby are examples of programming languages. HTML and XML are not programming languages, they are markup languages.

Progressive enhancement: Progressive enhancement is a strategy for web design that uses web technologies in a layered fashion that allows everyone to access the basic content and functionality of a web page, using any browser or Internet connection, while also providing those with better bandwidth or more advanced browser software an enhanced version of the page.

Property: Property is a CSS term and is roughly equivalent to an HTML tag. Properties are what define how a style should appear on a given web page.

Pseudo-class: Like pseudo-elements, pseudo-classes are used to add special effects to certain CSS selectors.

Pseudo-element: A pseudo-element is an element used to add a special effect to certain selectors.

Python: A sophisticated computer language that is commonly used for Internet applications. Designed to be a very readable language, it is named after Monty Python. Python files generally end in .py.

Reddit: A news aggregation website that is timely, interactive, and personalized. Users submit links to blog posts, photos, or videos and then other users vote those links up or down. The result is a list of the most engaging stories on the web at any given time.

Relational database: A piece of software that stores data in a series of tables, with relationships defined between them. A news story might have columns for a headline, date, text, and author, where the author points to another table containing the author's first name, last name, and email address. Information must be structured, but this allows for powerful queries. Examples include MySQL, Oracle, PostgreSQL, and SQLite.

Resolution: Refers to the physical number of pixels displayed on a screen (such as 1280 ×1024). Unlike in print, display resolution does not refer to the number of pixels or dots per inch on a computer screen, as this can be changed by changing the resolution of the screen (which, of course, does not change the physical size of the screen).

Retweet: A re-posting of someone else's Twitter post (Tweet).

RSS (Really Simple Syndication): A standard for websites to push their content to readers through Web formats to create regular updates through a "feed reader" or "RSS Reader." The symbol is generally an orange square with radiating white quarter circles.

Ruby: An increasingly popular programming language known for being powerful yet easy to write with. Originally introduced in 1995 by Yukihiro "Matz" Matsumoto, Ruby has gained increasing traction since 2005 because of the Ruby on Rails development framework, which can create websites quickly. Ruby is open source and is very popular for content-based sites. Ruby on Rails—A popular Web framework based on the Ruby programming language that makes common development tasks easier "out of the box."

RWD (responsive web design): When a website is designed and developed to work and look good on any device, regardless of screen size.

SaaS (Software as a Service): A pricing strategy and business model, where companies build a software solution, usually business-to-business, and charge a fixed monthly rate to access it on the Internet. It is a type of cloud computing.

Schema: Generally, a schema is an XML document used in place of a DTD to describe other XML documents.

Scribd: A document-sharing site that is often described as a "YouTube for documents" because it allows other sites to embed its content. It allows people to upload files and others to download in various formats.

Script: Generally refers to a portion of code on an HTML page that makes the page more dynamic and interactive. Scripts can be written in a variety of languages, including JavaScript.

Scripting language: A programming language designed to be easy to use for every day or administrative tasks. It may involve trade-offs such as sacrificing some performance for ease of programming. Popular scripting languages include PHP, Perl, Python, and Ruby.

Selector: In CSS, the selector is the item a style will be applied to.

SEM (search engine marketing): A type of marketing that involves raising a company or product's visibility in search engines by paying to have it appear in search results for a given word.

Semantic markup: In semantic markup, content is written within XHTML tags that offer context to what the content contains. Basic semantic markup refers to using items like header and paragraph tags, though semantic markup is also being used to provide much more useful context to web pages in an effort to make the web as a whole more semantic.

Semantic web: A vision of the web that is almost entirely machine readable, in which documents are published in languages that are designed specifically for data. It was first articulated by Tim Berners-Lee in 2001. In many implementations, tags would identify the information, such as <ADDRESS> or <DATE>.

Sender domain: If you receive an email from info@yourcompany.com, yourcompany.com is the sender domain.

Sentiment analysis: Analyzing posts, comments, tweets, and messages to determine the attitude of the speaker with respect to the topic, brand, or cause. The audience and messages can then be classified as either positive, negative, or neutral sentiment. Sentiment analysis helps marketers to understand how people feel about their product or brand and how they are responding to campaigns.

SEO (search engine optimization): A suite of techniques for improving how a website ranks on search engines such as Google. SEO is often divided into "white hat" techniques, which (to simplify) try to boost ranking by improving the quality of a website, and "black hat" techniques, which try to trick search engines into thinking a page is of higher quality than it actually is.

Server-side: Server-side refers to scripts run on a web server, as opposed to in a visitor's browser. Server-side scripts often take a bit longer to run than a client-side script, as each page must reload when an action is taken.

SGML: Stands for Standard Generalized Markup Language. SGML is a markup language used for defining the structure of a document. SGML is not mentioned very often, but it is the markup language that serves as the basis for both XML and HTML.

Skype: Skype is a VOIP (voice-over-IP) and instant messaging service. Users can call, message, or videoconference other users for free, or can pay to call landlines and mobile phone numbers.

Snapchat: A photo-messaging mobile app where users can take photos, record videos, add text and drawings, and send them to a controlled list of recipients. Snapchat's unique selling point (USP) is that messages or "snaps" self-destruct. Users set a time limit for how long recipients can view their snaps after which they will be hidden from the recipient's device and deleted from Snapchat's servers.

SOAP: Stands for Simple Object Access Protocol. SOAP is an XML-based protocol exchanging information across the Internet to allow an application on one site to access an application or database on another site.

Social influencer: An individual who creates or shares interesting or valuable content with their niche, loyal audience of social media followers that respect and trust their opinion. Social media influencers are important to brands as they are very effective at spreading online messages, they are more trusted than brands, and they are deemed more authentic.

Social media: A broad term referring to the wide swath of content creation and consumption that is enabled by the many-to-many distributed infrastructure of the Internet. Unlike legacy media, where the audience is usually on the receiving end of content creation, social media generally allows three stages of interaction with content: (1) producing, (2) consuming, and (3) sharing. Social media is incredibly broad and refers to blogging, wikis, video-sharing sites like YouTube, photo-sharing sites like Flickr, and social networking sites like Facebook and Twitter.

Social media manager: The person responsible for planning and developing social media and content strategies for a brand, cause, or topic.

Social media marketing: Social media marketing refers to the process of gaining traffic or attention through social media networking sites, such as Facebook, Twitter, and YouTube.

Social media monitoring: An active monitoring of social media channels for information about a company or organization. It allows marketing and social media managers to find insights into a brands' overall visibility on social media, measure the impact of campaigns, identify opportunities for engagement, assess competitor activity, and share of voice, and be alerted to impending crises.

Social media policy: A corporate code of conduct that provides guidelines for employees who post content on social media networking sites either as part of their job or as a private person. A social media policy should outline what employees are and are not allowed to say on behalf of the company, and how to respond to crises.

Specification: A specification or functional specification as it is sometimes known as a document that offers an explicit definition and requirements for a web service or technology and generally includes how the technology is meant to be used, along with the tags, elements, and any dependencies.

Sprout Social: A social media management platform that allows users and brand managers to update, post, schedule, and respond via multiple social media platforms, including for Facebook, Twitter, and Google+.

Structured thesaurus: A group of preferred terms created for editorial use to normalize and more effectively classify content. For example, the AP Stylebook is similar to (but includes more rules than) a structured thesaurus in that it gives writers preferred terms to use and standards to follow, so everyone following AP Style writes the word "website" the same way.

StumbleUpon: StumbleUpon is an online and mobile discovery engine that finds and recommends web content to its users. Registered users list their interests and then StumbleUpon recommends content it thinks they would like. If a user likes the content, they give it a thumbs up, if not, a thumbs down. StumbleUpon learns from those user actions to tailor the content it serves.

Subscriber: A subscriber is a person who allows a company to send him/her messages through email or other personal communication means. These subscribers are high value to publishers and businesses alike. Subscribers keep coming back!

Tag: A tag is a set of markup characters that are used around an element to indicate its start and end. Tags can also include HTML or other code to specify how that element should look or behave on the page (see also, HTML Tag).

Taxonomy: A hierarchical classification system. In the world of content, this can be a hierarchy of terms (generally called nodes or entities) that are used to classify the category or subject content belongs to as well as terms that are included in the content. In many cases, website navigation systems appear taxonomical in that users narrow down from broad top-level categories to the granular feature they want to see. An ontology is similar to a taxonomy in that it is also a classification system with nodes or entities, but it is more complex and flexible because ontologies allow for nonhierarchical relationships. While in a taxonomy, a node can be either a broader term or narrower term, in an ontology, nodes can be related in any way.

Template: A template is a file used to create a consistent design across a website. Templates are often used in conjunction with a CMS and contain both structural information about how a site should be set up, but also stylistic information about how the site should look.

Transparency: In the context of news and information, a term describing openness about information that has become increasingly popular. In many cases, it is used to refer to the transparency of government releasing data to journalists and to the public.

Trending topic: Trending topics are those topics on Twitter being discussed more than any others. They are listed to the left of the main column and are updated in real time.

Triple C principle: A systems development and implementation methodology that recommends hierarchical pursuit of systems integration through communication, cooperation, and coordination. See www.dejimodel.com.

Tumblr: Tumblr is a microblogging social networking site where registered users can post text, photos, quotes, links, music, and videos from their web browser, mobile device, desktop, or email to their completely customizable Tumblr page.

Tweet: A message posted using Twitter. It can be no more than 140 characters long and it can contain links to other websites, blogs, pictures, and videos.

Twitter: Twitter is a microblogging social network, where registered users can post short (140 characters or less) updates, known as tweets, about what they are doing, thinking, reading, eating, or many more subjects.

Twitterverse: Also known as twitosphere or twittersphere, it is the collective postings made by Twitter users. Used to describe Twitter as an ecosystem or universe.

UI (user interface): The part of a software application or website that users see and interact with, which takes into account the visual design and the structure of the program. While graphic design is an element of user interface design, it is only a portion of the consideration.

Unique users or visitors: A site's total number of users or visitors over a certain length of time. Accuracy depends on each user logging in with a unique username to access the site.

Unsubscribe: When someone permanently removes themselves from your email database.

URI (Uniform Resource Identifier): The way to identify the location for something on the Internet. It is most familiarly in "http:" form, but also encompasses "ftp:" or mailto:

URL (Uniform Resource Locator): Often used interchangeably with the "address" of a web page, such as http://hackshackers.com. All URLs are URIs, but not vice versa. While humans are familiar with URLs as a way to see web pages, computer programs often use URLs to pass each other machine-readable content, such as RSS feeds or Twitter information.

URL Shortener: An online tool that reduces the character length of a URL (uniform resource locator) or web address. Commonly used on social media platforms that impose character limits on users' posts, such as Twitter.

Usability: Usability refers to how easy it is for a novice user to utilize a system or navigate a website in its intended manner. Good usability means that elements such as systems navigation, content, images, and any interactive elements are easy to use, functioning the way they were intended, efficient, and effective for project management purposes or other operational needs without any special training requirements for users (Badiru, 2014). Usability is often used interchangeably with user-friendliness.

UX (user experience): Generally referring to the area of design that involves the holistic interaction a user has with a product or a service. It incorporates many disciplines, including engineering, graphic design, content creation, and psychology.

Valid: Valid web pages are those that return no errors based on the type of HTML/XHTML specified in the doctype declaration at the beginning of the file. The validator checks that the code used on the web page conforms to the specifications for that version of HTML/XHTML.

Vimeo: Vimeo is a video-sharing website on which users can upload, share, and view videos. Unlike YouTube, there are no advertisements before, during, or after your video.

Vine: Vine is a mobile video-sharing app, owned by Twitter, videos created with Vine have a maximum clip length of 7 seconds and play on a continuous loop.

Viral: Viral, in digital terms, comes from the medial term used to describe a small infections agent that can infect all types of organisms. "Going viral" online is when a piece of content spreads just like a virus, "infecting" people when they see it by evoking emotions that spur the viewer to share it.

Visit: A visit is a single user coming to a website. The same visitor returning multiple times means multiple visits.

Visitor: A visitor is a single user coming to a website. The same visitor returning multiple times in one day is still measured as just one visitor.

Web 2.0: Referring to the generation of Internet technologies that allow for interactivity and collaboration on websites. In contrast to Web 1.0 (roughly the first decade of the World Wide Web) where static content was downloaded into the browser and read, Web 2.0 uses the Internet as the platform. Technologies such as Ajax, which allow for rapid communication between the browser and the web server, underlie many Web 2.0 sites. The term was popularized by a 2004 conference, held by O'Reilly Media and MediaLive, called Web 2.0. (Also see Ajax)

Web 3.0: Sometimes used to refer to the semantic web. (Also see semantic web) webOS—Operating system used on the latest generation of Palm smartphones, including the Pre and the Pixi. Apps for webOS are developed using web standards (HTML, JavaScript, and CSS), which means there is a low barrier to entry for web developers to create mobile apps for webOS as compared to other mobile platforms. It allows for having several applications open at the same time, unlike the current iPhone.

Web page: A web page is a single document, generally written in HTML/ XHTML, meant to be viewed in a web browser. In many cases, web pages also include other coding and programming (such as PHP, Ruby on Rails, or ASP). Websites are generally built from multiple interlinked web pages.

Web server: A web server is a computer that has software installed and networking capabilities that allow it to host websites and pages and make them available to Internet users located elsewhere. There are various setups that can be used for a web server, including the LAMP setup previously mentioned.

Web standards: Standards are specifications recommended by the World Wide Web Consortium (W3C) for standardizing website design. The main purpose of web standards is to make it easier for both designers and those who create web browsers to make sites that will appear consistent across multiple platforms.

Webinar: A webinar is a seminar conducted over the Internet using video-conferencing, instant messaging, and VOIP services. Webinars are generally hosted by an organization or company and broadcast to a select group of individuals.

Widget: In a web context, this refers to a portable application that can be embedded into a third-party site by cutting and pasting snippets of code. Common web widgets include a Twitter box that can sit on a blog or a small Google Map that sits within an invitation. Desktop widgets, such as ones offered for the Macintosh Dashboard or by

Yahoo!, can be placed on the desktop of a computer, such as for weather or stocks.

Wiki: A website with pages that can be easily edited by visitors using their web browser, but generally now gaining acceptance as a prefix to mean "collaborative." Ward Cunningham created the first wiki, naming it WikiWikiWeb after the Hawaiian word for "quick." A wiki enables the audience to contribute to a knowledge base on a topic or share information within an organization, like a newsroom. The best-known wiki in existence is Wikipedia, which was introduced around 2000 as one of the first examples of mass collaborative information aggregation. Other sites that have been branded "wiki" include Wikinews, and Wikitravel. WikiLeaks has fallen out of favor in recent years.

WordPress: The most popular blogging software in use today, in large part because it is free and relatively powerful, yet easy to use.

XHTML: Stands for Extensible Hypertext Markup Language. Fundamentally, XHTML is HTML 4.0 that has been rewritten to comply with XML rules.

XML (Extensible Markup Language): A set of rules for encoding documents and data that goes beyond HTML capacities. Whereas HTML is generally concerned with the semantic structure of documents, XML allows other information to be defined and passed such as <vehicle>, <make>, <model>, <color> for a car.

REFERENCES

Badiru, A. B. (2014), editor, *Handbook of Industrial & Systems Engineering*, 2nd edition, Taylor & Francis Group/CRC Press, Boca Raton, FL.

Badiru, A. B. and LeeAnn Racz (2016), *Handbook of Measurements: Benchmarks for Systems Accuracy and Precision*, Taylor & Francis Group/CRC Press, Boca Raton, FL.

Badiru, A. B. and Marlin U. Thomas (2009), editors, *Handbook of Military Industrial Engineering*, Taylor & Francis Group/CRC Press, Boca Raton, FL.

Badiru, Adedeji B., Vhance C. Valencia and David Liu (2017), *Additive Manufacturing Handbook: Product Development for the Defense Industry*, Taylor & Francis Group/CRC Press, Boca Raton, FL.

Kerrigan, Laura (2021), "Your go-to glossary of digital terminology," Media Frontier online blog, https://www.mediafrontier.ch/blog/glossary-digital-terminology/, accessed March 28, 2021.

Index

Printed in the United States
by Baker & Taylor Publisher Services